Advances in Industrial Control

Series editors

Michael J. Grimble, Glasgow, UK
Michael A. Johnson, Kidlington, UK

More information about this series at http://www.springer.com/series/1412

John N. Jiang · Choon Yik Tang
Rama G. Ramakumar

Control and Operation of Grid-Connected Wind Farms

Major Issues, Contemporary Solutions, and Open Challenges

Springer

John N. Jiang
School of Electrical and Computer Engineering
University of Oklahoma
Norman, OK
USA

Rama G. Ramakumar
School of Electrical and Computer Engineering
Oklahoma State University
Stillwater, OK
USA

Choon Yik Tang
School of Electrical and Computer Engineering
University of Oklahoma
Norman, OK
USA

ISSN 1430-9491 ISSN 2193-1577 (electronic)
Advances in Industrial Control
ISBN 978-3-319-39133-5 ISBN 978-3-319-39135-9 (eBook)
DOI 10.1007/978-3-319-39135-9

Library of Congress Control Number: 2016939932

Printed on acid-free paper

This Springer imprint is published by Springer Nature
The registered company is Springer International Publishing AG Switzerland

To our families

Series Editors' Foreword

The series *Advances in Industrial Control* aims to report and encourage technology transfer in control engineering. The rapid development of control technology has an impact on all areas of the control discipline. New theory, new controllers, actuators, sensors, new industrial processes, computer methods, new applications, new philosophies…, new challenges. Much of this development work resides in industrial reports, feasibility study papers and the reports of advanced collaborative projects. The series offers an opportunity for researchers to present an extended exposition of such new work in all aspects of industrial control for wider and rapid dissemination.

The wind of change is sweeping through power generation and energy resource developments. Two significant drivers in this change are global concerns about the production of greenhouse gases from the use of coal and hydrocarbon fuels and a civilian nervousness about the safety and management of nuclear fuel use. Thus the renewable energy sources of hydro-power, solar energy, wind power, tidal power, and wave power are key areas for development. Several of them: hydro-, solar, wind and tidal are already a commercial reality and extending their penetration of the energy markets annually as new projects and schemes come to fruition. With differing characteristics these energy sources exhibit varying degrees of power quality, intermittency, and reliability. And while the modeling and control of individual devices has received and continues to receive research input from the control community, it is only in the last decade or so that the problem of integrating these renewable power resources into national and international power systems has become a topic of serious concern. This is partly due to the fact that it is only in recent years that large installations of renewable energy devices have come on stream. From the viewpoint of the grid, this can be seen as a "renewable energy electrical power station" to mirror the concept of a traditional electric power generating station.

This *Advances in Industrial Control* monograph *Control and Operation of Grid-Connected Wind Farms: Major Issues, Contemporary Solutions, and Open Challenges* by John N. Jiang, Choon Yik Tang, and Rama G. Ramakumar really begins from that novel viewpoint and examines the supervisory control of a group of wind turbines, that is a wind farm, to meet the operational requirements of the electrical power system. The monograph first reports the nonlinear control of a single grid-connected wind turbine. This provides a link to much existing wind turbine control system research and practice as can be found in several monographs in the *Advances in Industrial Control* series. The authors then progress to consider aspects of supervisory control of a grid-connected wind farm. An interesting feature of these chapters is the use of predictive, adaptive, and quasilinear control methods, from the toolbox of advanced control techniques, to solve the control design problems arising. The penultimate chapter of the monograph examines how the kinetic energy resource of a wind farm might be managed to meet grid operational considerations.

The authors of this monograph have excellent reputations in the wind energy and renewables energy field:

- John N. Jiang is Oklahoma Gas & Electric Co. Endowed Chair Professor in the School of Electrical and Computer Engineering at the University of Oklahoma and is the leader for the electric power engineering program at the university. Dr. Jiang has been involved in wind energy research since the early 1990s through a number of projects on technology research and development for the use of wind power. Currently Dr. Jiang is working with electric utility companies and research institutes on power system planning and operation taking into consideration the uncertainties and risks from integrating variable renewable energy resources, market-based demand responses, and stressed power system facilities.
- Choon Yik Tang is Associate Professor in the School of Electrical and Computer Engineering at the University of Oklahoma. Dr. Tang's current research interests include systems and control theory, distributed algorithms for computation and optimization over networks, and control and operation of wind farms.
- Rama G. Ramakumar is Regent Professor and PSO/Albrecht Naeter Professor and Director of the Engineering Energy Laboratory at Oklahoma State University. Over his 41-year career in electric power engineering, Dr. Ramakumar has been primarily involved in renewable energy and energy systems research. He is a Fellow of IEEE (1994) with a citation for "contributions to renewable energy systems and leadership in power engineering education."

This is an excellent addition to the *Advances in Industrial Control* series and a valuable complement to the existing contributions on the control of individual wind turbines.

Michael J. Grimble
Michael A. Johnson
Industrial Control Centre
Glasgow, Scotland, UK

Preface

The fundamental technical challenge in integrating wind-generated electrical energy into future power supply portfolios stems from mismatches between the stochastic nature of wind resources and the exigent requirements associated with power system reliability in terms of frequency, voltages, harmonics, security, ramp rates, etc. Thus, utilization of a meaningful amount of wind energy calls for advanced control strategies that take into account the requirements imposed by both sides of the interconnection.

This monograph aims at advancing the subject of wind energy control from the individual wind turbine level to the wind farm level. The ensemble of wind farm control technologies should enable reconfigurable, supervisory regulation of aggregated active and reactive power outputs from a large number of grid-connected wind turbines. The ultimate objective of such technologies would be to expand the capabilities of wind electric conversion systems by allowing them to coordinate with other energy resources, follow dispatch instructions from power grid operators, and adapt automatically to various conditions to alleviate the negative impact of wind irregularities on power system reliability and the quality of power supply. To this end, conventional single-turbine control technologies, though sufficient for individual turbine operation, may not be adequate at the wind farm level, especially when the level of wind energy penetration is high.

The monograph addresses a number of major issues in wind energy control in the context of power system operation. Following an overview of the required and desired objectives of grid integration and operation, the monograph devotes one chapter to each of the following topics: reconfigurable nonlinear control of single-turbine active and reactive power outputs; approximate model of wind turbine control systems for wind farm power control; contemporary techniques for designing and analyzing controllers at the wind farm level (two chapters); and control of kinetic energy stored in wind farms for transient frequency stability and the associated open challenges.

Wind farm control technologies are still at an early stage of development despite the rapid growth of wind power around the world. This monograph is primarily intended for control engineers/researchers who wish to apply their expertise to address key control issues that arise in the integration of wind farms into power grids. In addition, coverage of contemporary solutions to fundamental integration and operational problems in the monograph will benefit power engineers/ researchers who wish to promote wind energy with its multi-faceted benefits. We hope that the monograph will stimulate the interest of graduate students and collaboration between practicing engineers and academic researchers.

We are grateful to the National Science Foundation for supporting our research on wind energy control. We are also indebted to the University of Oklahoma and Oklahoma State University for providing environments that allow us to write this monograph. Furthermore, we would like to express our sincere gratitude to Emeritus Professor Michael A. Johnson of the University of Strathclyde. The monograph would not have been possible without his invitation and encouragement. Last but not least, we wish to thank Mr. Oliver Jackson and the staff at Springer for their patience and support during the writing of this monograph.

Norman, OK, USA John N. Jiang
Norman, OK, USA Choon Yik Tang
Stillwater, OK, USA Rama G. Ramakumar
January 2016

Contents

Chapter 1
Introduction

1.1 Background

The nexus among several economic, environmental, and geopolitical considerations has stimulated increasing and worldwide development of renewable energy resources for electric power generation. Setting aside the mature and well-known hydropower, wind electric conversion systems and photovoltaic systems are presently leading the list of renewable energy technologies. As their economic viability grows, their penetration into conventional electric utility systems is expected to increase, ushering in a series of technical issues in the realm of reliability, power quality, controllability, and commitment, among others, due to the variable nature of their power outputs.

Key among these technical issues is the development of control strategies for wind electric conversion in terms of the so-called real power or active power, which is the power directly converted from wind. Although the ultimate goal is to properly control both the active power and reactive power under various grid operating conditions, active power control is of greater concern in the integration of wind energy into power grids. The main reason is that active power directly affects the economic merit and wide-area dynamic stability of a power system, where electricity is produced from wind, transported through the system, and consumed by customers. In other words, active power related problems generally have a broader, system-wide impact, making them more challenging. In contrast, most reactive power related problems are local issues pertaining to the adequacy of the so-called ancillary services. With that being said, reactive power control is also a vital aspect of wind energy control, which supports the electromagnetic field and local voltages required for conversion and transportation of active power. Note that although these two types of power are electromagnetically-coupled duals in principle, they can often be studied separately, and treated independently for simplicity.

Traditionally, control of active power is accomplished by blade feathering and torque control, in which the blade pitch angles and electromagnetic torques are adjusted to extract a maximum amount of available wind energy. To realize this maximum power tracking objective, various types of single-turbine control tech-

J.N. Jiang et al., *Control and Operation of Grid-Connected Wind Farms*,
Advances in Industrial Control, DOI 10.1007/978-3-319-39135-9_1

nologies have been developed and implemented. Indeed, some of these technologies also attempt to maximize turbine lifetime by reducing mechanical load, vibration, and fatigue which are primarily affected by active power. At present, single-turbine active power control technologies have developed to the point of being able to meet basic integration requirements imposed by power grids, including: variable-speed constant-frequency control of active power, efficient energy conversion, fast and effective power electronic converters, and minimization of mechanical load, vibration, and fatigue. In fact, many of these technologies are commercially available today.

Despite the aforementioned achievement, advanced techniques for active power control are still needed to better understand and manage the negative impact of wind energy on power system reliability. Indeed, with the grid integration requirements becoming increasingly rigid, and with new ideas on wind energy control continually being discovered, further advances in active power control are in pressing demand. Such requirements and ideas include, but are not limited to, grid frequency control, amount and quality of wind power, coordinated operation with other energy resources, and smart control capabilities under different grid operating and electricity market conditions.

In short, advances in active power control for wind electric conversion systems are essential for improving existing control solutions to classic problems, creating new capabilities, and ultimately enhancing the level of acceptance of wind energy penetration into power grids.

1.2 State of the Art in Wind Farm Power Control

The frontier of research in active power control for wind electric conversion systems is expanding from single-turbine considerations to farm level considerations. This is because utilization of a meaningful amount of wind energy relies heavily on electric power generated by wind farms. With the installation of more wind farms and with their aggregated power outputs becoming a significant portion of the total generation in power grids, it is increasingly evident that the performance of individual turbines matters much less than the overall performance of wind farms. Moreover, recent research has shown that the electrical and mechanical characteristics of a large collection of wind electric conversion systems may be very different from those of individual ones, especially during dynamic and transient periods. These observations motivate the development of sophisticated wind farm controllers that can deal with hundreds or thousands of turbines, so that wind energy can be properly harvested, seamlessly coordinated with other energy resources, and suitably adapted to grid operating conditions. The observations also imply that existing single-turbine-based knowledge may not be adequate for understanding the dynamic interactions between wind farms and power grids, let alone their true impact on grid security and reliability.

An advanced wind farm controller can be seen as a supervisory control scheme that optimizes the performance of a wind farm by sending appropriate control signals to individual turbines. The performance may be measured in several ways, most notably the amount and quality of wind farm power output, the amount of reduction in mechanical load, and the ability to adapt to grid operating and electricity market conditions. Thus, to perform well with respect to these measures, a supervisory control scheme should take into account factors at the wind farm level—including wind availability and correlation, wind farm related variables such as the turbine layout and types, and grid related variables such as the grid operating conditions and status of other energy resources—in addition to feedback at the turbine level.

Historically, active power control for wind electric conversion systems has followed a decentralized approach: most control schemes were designed around a single turbine to maximize its own energy conversion and/or minimize its own mechanical load, irrespective of what other nearby turbines are doing. While this approach is simple, it offers significant room for improvement because when a set of grid-connected turbines are deployed over a geographical area sharing a wind field, they tend to interact electrically through the grid, and aerodynamically through the wind. Such interactions may dramatically affect the overall wind farm performance, but are completely disregarded by a decentralized control scheme. Hence, it is conceivable that such a scheme may be substantially improved by adding a centralized, supervisory unit that coordinates the turbines, forcing them to cooperate.

Improvements in wind farm control technologies could have notable economic and engineering ramifications. From the economic perspective, such improvements could increase the ability of power grids in accepting more variable and intermittent wind energy without incurring serious costs in the form of spilled power, energy storage, and procurement of an excessive amount of expensive spinning reserve. From the engineering perspective, such improvements not only could enhance the accuracy and smoothness of wind farm power outputs, they also could turn wind farms into a source of ancillary services, providing power grids with, say, primary and secondary frequency support during power system contingencies.

To date, a number of research and industrial projects have been carried out to gain a better understanding of the potential of wind farm power control. Some of these projects focus on input factors such as the dynamics and energy distribution of a wind field, while others aim at integration issues such as coordination with other energy resources and adaptation to grid operating conditions. For instance, the National Renewable Energy Laboratory (NREL) in the United States recently initiated a research project to develop wind forecast models and modeling tools that account for atmospheric turbulences and wake effects [4]. These models and tools are expected to provide wind farm operators and power grid operators with more accurate information on the characteristics of incoming wind, which are useful for determining appropriate control signals for individual turbines. As another example, the Aeolus research project on distributed control of large-scale offshore wind farms was recently completed in Europe [55, 63]. Among the key project outcomes are studies on wind farm/speed modeling and simulation [64, 100, 102] and wind farm control [101, 103, 104]. Yet another example is the wind farm management system called WindCONTROL of General Electric (GE) [30], which aims at delivering an intelligent solution to reactive power management in wind farms, so that the quality of voltages can be maintained in the presence of active power fluctuations induced by wind.

Despite these research efforts in the wind industry, development of advanced wind farm control technologies is still at an early stage, especially those that deal with active power control. As mentioned previously, active power control is important because it directly affects the performance of a wind farm and the wide-area dynamic stability of a power system. In addition, it is challenging because it involves many grid related variables such as the grid operating conditions, as well as many local variables such as the wind farm composition and wind field dynamics. Because it is important and challenging, wind farm active power control has emerged as a cutting-edge research direction in the control and power system engineering communities.

1.3 Scope of Monograph

This monograph addresses a number of major issues that arise in the control and operation of grid-connected wind farms. More specifically, we show how systems and control theory may be applied to develop contemporary solutions to important issues such as modeling of wind farms with numerous wind turbines for control purposes, and control of wind farm active power outputs with multiple operational objectives and constraints.

Over the past decades, various technologies and control algorithms for individual wind electric conversion systems have been developed and documented in numerous articles, reports, and books. As many of these publications have dealt in great detail with topics such as rotating machines, solid-state converters, single-turbine control algorithms, and extension of turbine lifetime, these topics are not within the scope of this monograph. Moreover, as no single book can satisfy the needs of all readers, the monograph is geared toward readers from the control and power system engineering communities.

The outline of this monograph is as follows: we first introduce a single-turbine model and a dual-mode controller that set the stage for the rest of the monograph. We then construct and validate a generic approximate model that can represent different types of wind electric conversion systems for the purpose of research in wind farm power control. Based on this approximate model, we next develop a supervisory wind farm controller, which enables turbines in a wind farm to cooperate in a number of ways. The ability to cooperate, in turn, expands the capabilities of the wind farm, allowing it to accurately follow dispatch instructions from power grid operators and better adapt to grid operating and electricity market conditions. These new capabilities are key to alleviating the negative impact of wind variability and intermittency on power system reliability, in addition to promoting the economic viability of wind. Finally, we address a very new and somewhat controversial topic in wind farm active power control, namely, the topic of controlling the kinetic energy stored in wind farms as a means to maintain transient frequency stability during a power imbalance.

In terms of the monograph chapters, Chap. 2 describes a mathematical model for a variable-speed wind turbine with a doubly fed induction generator. It also uses the model to design a reconfigurable nonlinear controller that enables the turbine to operate in both the maximum power tracking and power regulation modes, with the latter mode being critical in expanding the capabilities of wind farms. Chapter 3 develops a simple approximate model that closely mimics the active and reactive power dynamics of various wind turbine control systems. It also shows that the model facilitates the design and analysis of wind farm controllers. Chapter 4 introduces a novel supervisory wind farm controller consisting of a model predictive controller and an adaptive controller, which enables the active power output of a wind farm to accurately and smoothly track a desired reference from a power grid operator. Chapter 5 shows that the recently-developed theory of Quasilinear Control can be utilized to optimize the gains of the said wind farm controller, significantly improving its tracking performance over a broad range of operating regimes. Chapter 6 investigates the topic of kinetic energy release in wind farms from a theoretical standpoint and provides answers to the fundamental question of what can, and cannot, be achieved with kinetic energy release. Finally, Chap. 7 concludes the monograph.

Chapter 2
Single Wind Turbine Power Generation Systems

Abstract For a power system with substantial power generation from wind farms, controllability of the wind farm power outputs is critical to power system reliability and economy. Both the active and reactive powers need to be maintained at appropriate levels. Indeed, recent experience with wind farm operation and research suggests that a wind farm should have at least two operating modes: *maximum power tracking* (MPT) and *power regulation* (PR). MPT is a traditional operating mode, aimed at enabling wind turbines in a wind farm to convert as much of the energy in wind to electrical energy as possible under normal operation conditions. PR, on the other hand, is concerned with adjusting the wind turbine power outputs as needed by power system reliability, or economic conditions. Being able to operate in either the MPT or PR mode is becoming increasingly important as the penetration of wind energy increases. In this chapter, we first introduce the basic structure of and a mathematical model for a variable-speed wind turbine with a doubly fed induction generator (DFIG), a widely used power generation technology today. This model may be used in the development of controllers for controlling the active and reactive power outputs of the wind turbine. Indeed, we also illustrate an application of the model to the design of a reconfigurable nonlinear controller, which enables the wind turbine to maximize its active power in the MPT mode, regulate its active power in the PR mode, switch between the two modes, and adjust its reactive power to achieve a desired power factor, while coping with uncertainties in most of its parameters. Finally, we demonstrate the effectiveness of the controller through simulation with a realistic wind profile.

2.1 Introduction

The basic structure of a typical variable-speed wind turbine with a DFIG is shown in Fig. 2.1. The DFIG is an electric generator employed in most variable-speed wind turbines, which is a three-phase slip-ring induction machine, with its stator windings directly connected to the power grid, and its rotor windings connected to the grid through a bidirectional four-quadrant power electronic converter which can vary the frequency, magnitude, and phases of the voltage, or current, in the rotor windings. The

J.N. Jiang et al., *Control and Operation of Grid-Connected Wind Farms*,
Advances in Industrial Control, DOI 10.1007/978-3-319-39135-9_2

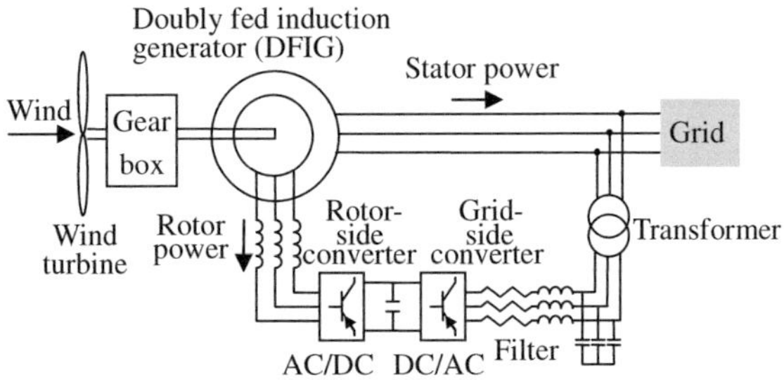

Fig. 2.1 Basic structure of a typical variable-speed wind turbine with a DFIG

wind turbine is of a variable blade pitch angle type, whose aerodynamic properties are characterized by a family of performance coefficient curves also known as C_p-curves or C_p-surface. Typically, a fixed ratio gearbox is used to couple the wind turbine and the DFIG. The gearbox, however, is usually ignored in many studies on wind turbine control, and this monograph is no exception.

In general, both the stator and rotor windings are three-phase circuits. In this monograph, the analysis of two three-phase circuits, including the electrical variables such as voltages, currents, and impedances, will be carried out using the synchronously rotating direct-quadrature axis (dq-axis) frame, where the relationship between three-phase quantities and dq components is given in terms of the direct–quadrature–zero (or $dq0$) transformation and its inverse [15]. In a typical implementation for balanced three-phase circuits of stators and rotors, the dq analysis is based on a synchronously rotating dq-axis frame, where the d-axis is oriented along the stator-flux vector position. The study can then be carried out and implemented using vector control in the stator-flux orientation to achieve a decoupled control between electrical torque and rotor excitation voltage or current. In the dq-axis frame, independent control of the active and reactive power outputs can be achieved, just as in the case of a synchronous generator, so the analysis and design of controllers can be carried out succinctly without losing any generality.

Furthermore, in the analysis and design of control algorithms, we assume that the rotor-side circuit, including the rotor-side converter, grid-side convertor, capacitor, and transformer as shown in Fig. 2.1, are ideal. We also assume that any control algorithm for the voltages and currents of the rotor windings can be physically realized.

2.2 Mathematical Model

2.2.1 Electrical Dynamics

Consider a variable-speed wind turbine with a DFIG, as shown in Fig. 2.1. For this wind turbine, the dynamics of its electrical part can be represented by a fourth-order state space model. To see this, note that the voltage equations are [38]

$$v_{ds} = R_s i_{ds} - \omega_s \varphi_{qs} + \dot{\varphi}_{ds}, \quad (2.1)$$
$$v_{qs} = R_s i_{qs} + \omega_s \varphi_{ds} + \dot{\varphi}_{qs}, \quad (2.2)$$
$$v_{dr} = R_r i_{dr} - (\omega_s - \omega_r)\varphi_{qr} + \dot{\varphi}_{dr}, \quad (2.3)$$
$$v_{qr} = R_r i_{qr} + (\omega_s - \omega_r)\varphi_{dr} + \dot{\varphi}_{qr}, \quad (2.4)$$

where $v_{ds}, v_{qs}, v_{dr}, v_{qr} \in \mathbb{R}$ are the d- and q-axis components of the stator and rotor voltages; $i_{ds}, i_{qs}, i_{dr}, i_{qr} \in \mathbb{R}$ are the d- and q-axis components of the stator and rotor currents; $\varphi_{ds}, \varphi_{qs}, \varphi_{dr}, \varphi_{qr} \in \mathbb{R}$ are the d- and q-axis components of the stator and rotor fluxes; $\omega_s > 0$ is the constant angular velocity of the synchronously rotating reference frame; $\omega_r > 0$ is the rotor angular velocity; and R_s, R_r are the stator and rotor resistances. The flux equations are [38]

$$\varphi_{ds} = L_s i_{ds} + L_m i_{dr}, \quad (2.5)$$
$$\varphi_{qs} = L_s i_{qs} + L_m i_{qr}, \quad (2.6)$$
$$\varphi_{dr} = L_m i_{ds} + L_r i_{dr}, \quad (2.7)$$
$$\varphi_{qr} = L_m i_{qs} + L_r i_{qr}, \quad (2.8)$$

where L_s, L_r, and L_m are the stator, rotor, and mutual inductances, respectively, satisfying $L_s > L_m$ and $L_r > L_m$. From (2.5)–(2.8), the current equations can be written as

$$i_{ds} = \frac{1}{\sigma L_s}\varphi_{ds} - \frac{L_m}{\sigma L_s L_r}\varphi_{dr}, \quad (2.9)$$
$$i_{qs} = \frac{1}{\sigma L_s}\varphi_{qs} - \frac{L_m}{\sigma L_s L_r}\varphi_{qr}, \quad (2.10)$$
$$i_{dr} = -\frac{L_m}{\sigma L_s L_r}\varphi_{ds} + \frac{1}{\sigma L_r}\varphi_{dr}, \quad (2.11)$$
$$i_{qr} = -\frac{L_m}{\sigma L_s L_r}\varphi_{qs} + \frac{1}{\sigma L_r}\varphi_{qr}, \quad (2.12)$$

where $\sigma = 1 - L_m^2/(L_s L_r)$ is the leakage coefficient. Selecting the four fluxes as state variables and substituting (2.9)–(2.12) into (2.1)–(2.4), the electrical dynamics in state space form can be written as

$$\dot{\varphi}_{ds} = -\frac{R_s}{\sigma L_s}\varphi_{ds} + \omega_s \varphi_{qs} + \frac{R_s L_m}{\sigma L_s L_r}\varphi_{dr} + v_{ds}, \quad (2.13)$$
$$\dot{\varphi}_{qs} = -\omega_s \varphi_{ds} - \frac{R_s}{\sigma L_s}\varphi_{qs} + \frac{R_s L_m}{\sigma L_s L_r}\varphi_{qr} + v_{qs}, \quad (2.14)$$
$$\dot{\varphi}_{dr} = \frac{R_r L_m}{\sigma L_s L_r}\varphi_{ds} - \frac{R_r}{\sigma L_r}\varphi_{dr} + (\omega_s - \omega_r)\varphi_{qr} + v_{dr}, \quad (2.15)$$
$$\dot{\varphi}_{qr} = \frac{R_r L_m}{\sigma L_s L_r}\varphi_{qs} - (\omega_s - \omega_r)\varphi_{dr} - \frac{R_r}{\sigma L_r}\varphi_{qr} + v_{qr}. \quad (2.16)$$

Treating the rotor voltages v_{dr} and v_{qr} as control variables and the stator voltages v_{ds} and v_{qs} as constants (that are not simultaneously zero), the dynamics (2.13)–(2.16) can be written in a matrix form as

$$\begin{bmatrix} \dot{\varphi}_{ds} \\ \dot{\varphi}_{qs} \\ \dot{\varphi}_{dr} \\ \dot{\varphi}_{qr} \end{bmatrix} = \underbrace{\begin{bmatrix} -\frac{R_s}{\sigma L_s} & \omega_s & \frac{R_s L_m}{\sigma L_s L_r} & 0 \\ -\omega_s & -\frac{R_s}{\sigma L_s} & 0 & \frac{R_s L_m}{\sigma L_s L_r} \\ \frac{R_r L_m}{\sigma L_s L_r} & 0 & -\frac{R_r}{\sigma L_r} & \omega_s \\ 0 & \frac{R_r L_m}{\sigma L_s L_r} & -\omega_s & -\frac{R_r}{\sigma L_r} \end{bmatrix}}_{A} \begin{bmatrix} \varphi_{ds} \\ \varphi_{qs} \\ \varphi_{dr} \\ \varphi_{qr} \end{bmatrix} + \underbrace{\begin{bmatrix} 0 & 0 \\ 0 & 0 \\ 1 & 0 \\ 0 & 1 \end{bmatrix}}_{B} \begin{bmatrix} v_{dr} \\ v_{qr} \end{bmatrix} + \begin{bmatrix} v_{ds} \\ v_{qs} \\ -\omega_r \varphi_{qr} \\ \omega_r \varphi_{dr} \end{bmatrix}, \tag{2.17}$$

where A and B are constant matrices as defined in (2.17). Similarly, the flux–current relationship (2.5)–(2.8) can be written in a matrix form as

$$\underbrace{\begin{bmatrix} \varphi_{ds} \\ \varphi_{qs} \\ \varphi_{dr} \\ \varphi_{qr} \end{bmatrix}}_{\varphi} = \begin{bmatrix} L_s & 0 & L_m & 0 \\ 0 & L_s & 0 & L_m \\ L_m & 0 & L_r & 0 \\ 0 & L_m & 0 & L_r \end{bmatrix} \underbrace{\begin{bmatrix} i_{ds} \\ i_{qs} \\ i_{dr} \\ i_{qr} \end{bmatrix}}_{i}, \tag{2.18}$$

where φ and i are vectors as defined in (2.18).

Neglecting power losses associated with stator and rotor resistances, the active and reactive stator and rotor powers are given by [68]

$$P_s = -v_{ds} i_{ds} - v_{qs} i_{qs}, \tag{2.19}$$
$$Q_s = -v_{qs} i_{ds} + v_{ds} i_{qs}, \tag{2.20}$$
$$P_r = -v_{dr} i_{dr} - v_{qr} i_{qr}, \tag{2.21}$$
$$Q_r = -v_{qr} i_{dr} + v_{dr} i_{qr}, \tag{2.22}$$

and the total active and reactive powers of the turbine are

$$P = P_s + P_r, \tag{2.23}$$
$$Q = Q_s + Q_r, \tag{2.24}$$

where positive (negative) values of P and Q mean that the turbine injects power into (draws power from) the power grid.

2.2.2 Mechanical Dynamics

The dynamics of the mechanical part of the wind turbine can be represented by a first-order state space model [15]

$$J\dot{\omega}_r = T_m - T_e - C_f\omega_r, \tag{2.25}$$

where the rotor angular velocity ω_r is another state variable, J is the moment of inertia, C_f is the friction coefficient, T_m is the mechanical torque generated, and T_e is the electromagnetic torque which can be expressed as [68]

$$T_e = \varphi_{qs}i_{ds} - \varphi_{ds}i_{qs}, \tag{2.26}$$

where a positive (negative) value of T_e corresponds to the turbine acting as a generator (motor). It is known that [14] the mechanical power P_m converted from wind through the turbine blades is given by

$$P_m = T_m\omega_r = \frac{1}{2}\rho AC_p(\lambda, \beta)V_w^3, \tag{2.27}$$

where ρ is the air density, $A = \pi R^2$ is the area swept by the rotor blades of radius R, V_w is the wind speed, and $C_p(\lambda, \beta)$, commonly referred to as the C_p-surface, is the performance coefficient of the wind turbine, whose value is a function of the tip speed ratio $\lambda \in (0, \infty)$, defined as

$$\lambda = \frac{\omega_r R}{V_w}, \tag{2.28}$$

and the blade pitch angle $\beta \in [\beta_{\min}, \beta_{\max}]$, which is another control variable.

The performance coefficient $C_p(\lambda, \beta)$ is typically provided by turbine manufacturers and may vary greatly from one turbine to another [14]. Therefore, to make the results of this chapter broadly applicable to a wide variety of turbines, no specific expression of $C_p(\lambda, \beta)$ will be assumed. Instead, $C_p(\lambda, \beta)$ will only be assumed to satisfy the following mild assumptions for the purpose of analysis:

A1. Function $C_p(\lambda, \beta)$ is continuously differentiable in both λ and β over $\lambda \in (0, \infty)$ and $\beta \in [\beta_{\min}, \beta_{\max}]$.

A2. There exists $c \in (0, \infty)$ such that for all $\lambda \in (0, \infty)$ and $\beta \in [\beta_{\min}, \beta_{\max}]$, we have $C_p(\lambda, \beta) \le c\lambda$. This assumption is mild because it is equivalent to saying that the mechanical torque T_m is bounded from above, since $T_m \propto \frac{C_p(\lambda,\beta)}{\lambda}$ according to (2.27) and (2.28).

A3. For each fixed $\beta \in [\beta_{\min}, \beta_{\max}]$, there exists $\lambda_1 \in (0, \infty)$ such that for all $\lambda \in (0, \lambda_1)$, we have $C_p(\lambda, \beta) > 0$. This assumption is also mild because turbines are designed to capture wind power over a wide range of λ, including times when λ is small.

A4. There exist $\underline{c} \in (-\infty, 0)$ and $\overline{c} \in (0, \infty)$ such that for all $\lambda \in (0, \infty)$ and $\beta \in [\beta_{\min}, \beta_{\max}]$, we have $\underline{c} \le \frac{\partial}{\partial\lambda}(\frac{C_p(\lambda,\beta)}{\lambda}) \le \overline{c}$.

As it follows from the above, the wind turbine is modeled as a fifth-order nonlinear dynamical system with state variables φ_{ds}, φ_{qs}, φ_{dr}, φ_{qr}, and ω_r; control variables v_{dr}, v_{qr}, and β; output variables P and Q; exogenous "disturbance" V_w;

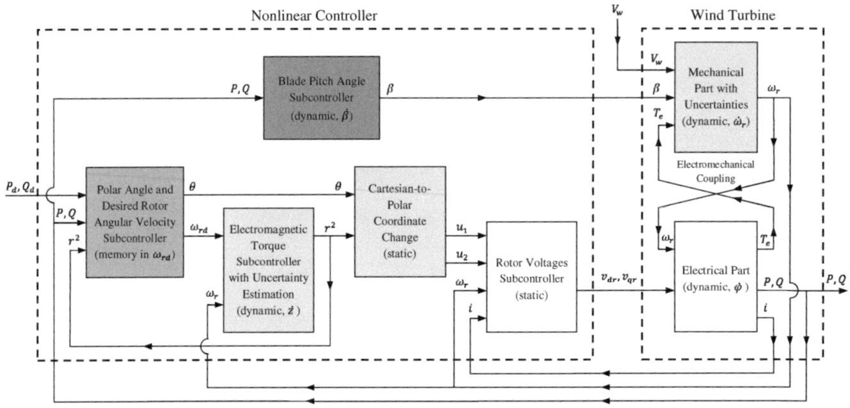

Fig. 2.2 Block diagram of the wind turbine and the nonlinear dual-mode controller

nonlinear state equations (2.17) and (2.25); and nonlinear output equations (2.19)–(2.24). Notice that the system dynamics are strongly coupled: the mechanical state variable ω_r affects the electrical dynamics bilinearly via the third and fourth rows of (2.17) (i.e., (2.15) and (2.16)), while the electrical state variables φ_{ds}, φ_{qs}, φ_{dr}, and φ_{qr} affect the mechanical dynamics quadratically via (2.9)–(2.12), (2.25), and (2.26). Since the stator winding of the DFIG is directly connected to the grid, for reliability reasons the stator voltages v_{ds} and v_{qs} are assumed to be fixed and not to be controlled. Moreover, since (2.18) represents a bijective mapping between vectors φ and i and since the currents i and the rotor angular velocity ω_r can be measured, a controller for this system has access to all of its states (i.e., full state feedback is available). A block diagram of this system is shown on the right-hand side of Fig. 2.2, in which the electromechanical coupling can be seen.

In the rest of the chapter, we will illustrate an application of this wind turbine model to the design of a reconfigurable nonlinear controller. As will be shown, this controller not only allows the turbine to operate in either the MPT or PR mode by controlling its active and reactive powers via its rotor voltages and blade pitch angle, but also allows the turbine to have uncertainties in most of its parameters.

2.3 Nonlinear Dual-Mode Control with Uncertainties

Most grid-connected large-scale wind farms today operate in the *maximum power tracking* (MPT) mode, making their wind turbines harvest as much wind energy as possible, following the "let it be when the wind blows" philosophy of operation. With the current, relatively low level of wind energy penetration in the power generation portfolio, this MPT mode of operation does not cause significant issues. However, with the anticipated increase in penetration in the near future [2], MPT can negatively impact power system reliability, such as producing excessive power that destabilizes a grid [1]. Thus, it is important that a wind farm can also operate in the so-called *power regulation* (PR) mode, whereby the total power output from its wind turbines is closely regulated at a desired setpoint, despite the fluctuating wind.

The ability of a wind farm to operate in either the MPT or PR mode, and switch seamlessly between them, is highly beneficial: not only does the PR mode provide a cushion to absorb the impact of wind fluctuations on the total power output, it also enables a power system to effectively respond to changes in reliability conditions and economic signals. For instance, when a sudden drop in load occurs, the power system may ask the wind farm to switch from MPT to PR and generate less power, rather than rely on expensive down-regulation generation. As another example, the PR mode, when properly designed, allows the total power output to smoothly and accurately follow system dispatch requests, thus reducing the reliance on ancillary services such as reliability reserves.

In the next section, we will develop a wind turbine controller with which a wind farm can have the aforementioned ability. More specifically, we will use the model from Sect. 2.2 to design a nonlinear dual-mode controller that controls the rotor voltages and blade pitch angle of a wind turbine, so that the turbine is capable of maximizing its active power in the MPT mode, regulating its active power in the PR mode, switching seamlessly between the two modes, adjusting its reactive power to achieve a desired power factor, and coping with inevitable uncertainties in most of its aerodynamic and mechanical parameters. The controller design will consist of the following steps: first, we show that although the dynamics of a wind turbine are highly nonlinear and electromechanically coupled, they possess a structure that makes the electrical part feedback linearizable, so that arbitrary pole placement can be carried out. We also show that because the electrical dynamics can be made very fast, it is possible to perform model order reduction, so that only the first-order mechanical dynamics remain to be considered. Next, we show that parametric uncertainties in the mechanical dynamics can be lumped and estimated via an uncertainty estimator which, together with a torque controller, enables the rotor angular velocity to track a desired reference whenever possible. Finally, we introduce a potential function that measures the difference between the actual and desired powers and present a gradient-like approach for minimizing this function.

We note that the current literature offers a large collection of wind turbine controllers, including [12, 27, 42, 43, 49, 52, 57–59, 65, 75, 77, 82, 87, 88, 95, 105, 108, 112–114]. However, most of the existing work consider the mechanical and electrical parts separately (e.g., [12, 27, 42, 43, 57–59, 82, 95, 105, 112] consider only the former, while [52, 65, 75, 77, 87, 88, 108, 113, 114] consider only the latter), and for a few of those (e.g., [49]) that consider both parts, its controller can only operate in the MPT mode as opposed to both the MPT and PR modes. Moreover, although the existing work has provided valuable understanding in the control of wind turbines, only a few address the issue of uncertainties. For example, [57–59] propose adaptive frameworks for controlling the mechanical part of wind turbines, so that the power captured is maximized, despite not knowing the turbine performance coefficient.

2.4 Controller Design

Given the model from Sect. 2.2, we address in this section the following problem: design a feedback controller that adjusts the rotor voltages v_{dr} and v_{qr} and blade pitch angle β, so that the active and reactive powers P and Q track—as closely as possible and limited only by wind strength—some time-varying desired references P_d and Q_d, presumably provided by a wind farm operator. When P_d is larger than what the wind turbine is capable of generating, it means that the operator wants the turbine to operate in the MPT mode; otherwise, the PR mode is sought. By also providing Q_d, the operator indirectly specifies a desired power factor $\text{PF}_d = P_d/\sqrt{P_d^2 + Q_d^2}$, around which the actual power factor $\text{PF} = P/\sqrt{P^2 + Q^2}$ should be regulated. The controller may use i, ω_r, P, and Q, which are all measurable, as feedback. The fluxes φ may also be viewed as feedback, since they are bijectively related to i through (2.18). Moreover, the controller may use values of all the electrical parameters (i.e., ω_s, R_s, R_r, L_s, L_r, L_m, v_{ds}, and v_{qs}) and turbine-geometry-dependent parameters (i.e., J, A, R, $\beta_{\min}$, and $\beta_{\max}$), since these values are typically quite accurately known. However, it may not use values of the C_p-surface, the air density ρ, and the friction coefficient C_f, since these values are inherently uncertain and can change over time. Furthermore, the controller should not rely on the wind speed V_w, since it may not be accurately measured.

Our solution to the above problem is a controller consisting of four subcontrollers. Figure 2.2 shows the architecture of this controller, in which the blocks represent its subcontrollers. Note that the controller accepts P_d and Q_d as reference inputs, uses i, ω_r, P, and Q as feedback, and produces v_{dr}, v_{qr}, and β as control inputs to the wind turbine. Moreover, the different gray levels of the blocks in Fig. 2.2 represent our intended timescale separation in the closed-loop dynamics: the darker a block, the slower its dynamics. The subcontrollers will be described in Sects. 2.4.1–2.4.4.

2.4.1 *Rotor Voltages Subcontroller*

Observe that although the electrical dynamics (2.17) are nonlinear, they possess a nice structure: the first and second rows of (2.17) are affine, consisting of linear terms and the constants v_{ds} and v_{qs}, while the third and fourth are nonlinear, consisting of linear terms, the control variables v_{dr} and v_{qr}, and the nonlinearities $-\omega_r\varphi_{qr}$ and $\omega_r\varphi_{dr}$ induced by the electromechanical coupling. Since the nonlinearities enter the dynamics the same way the control variables v_{dr} and v_{qr} do, we may use *feedback linearization* [62] to cancel them and perform *pole placement* [28], i.e., let

$$v_{dr} = \omega_r\varphi_{qr} - K_1^T\varphi + u_1, \tag{2.29}$$

$$v_{qr} = -\omega_r\varphi_{dr} - K_2^T\varphi + u_2, \tag{2.30}$$

where $\omega_r \varphi_{qr}$ and $-\omega_r \varphi_{dr}$ are intended to cancel the nonlinearities, $-K_1^T \varphi$ and $-K_2^T \varphi$ with $K_1, K_2 \in \mathbb{R}^4$ are for pole placement, and u_1 and u_2 are new control variables to be designed in Sect. 2.4.2.

Substituting (2.29) and (2.30) into (2.17), we obtain

$$\dot{\varphi} = (A - BK)\varphi + \begin{bmatrix} v_{ds} & v_{qs} & u_1 & u_2 \end{bmatrix}^T, \tag{2.31}$$

where $K = [K_1 \; K_2]^T$ is the state feedback gain matrix. Since the electrical dynamics are physically allowed to be much faster than the mechanicals, we may choose K in (2.31) to be such that $A - BK$ is asymptotically stable with very fast eigenvalues. With K chosen as such and with relatively slow-varying u_1 and u_2, the linear differential equation (2.31) may be approximated by a linear algebraic equation

$$\varphi = -(A - BK)^{-1} \begin{bmatrix} v_{ds} & v_{qs} & u_1 & u_2 \end{bmatrix}^T. \tag{2.32}$$

Consequently, the fifth-order state equations (2.17) and (2.25) may be approximated by the first-order state equation (2.25) along with algebraic relationships (2.29), (2.30), and (2.32). This approximation will be made in all subsequent development (but not in simulation).

Note that (2.18), (2.29), and (2.30) describe the *Rotor Voltages Subcontroller* block in Fig. 2.2.

2.4.2 Electromagnetic Torque Subcontroller with Uncertainty Estimation

Having addressed the electrical dynamics, we now consider the mechanicals, where the goal is to construct a subcontroller, which makes the rotor angular velocity ω_r track a desired, slow-varying reference ω_{rd}, despite not knowing most of the aerodynamic and mechanical parameters listed earlier.

To come up with such a subcontroller, we first introduce a coordinate change. As shown in [106], because of (2.18), (2.26), and (2.32), the electromagnetic torque T_e may be expressed as a quadratic function of the new control variables u_1 and u_2, i.e.,

$$T_e = \begin{bmatrix} u_1 & u_2 \end{bmatrix} \begin{bmatrix} q_1 & q_2 \\ q_2 & q_3 \end{bmatrix} \begin{bmatrix} u_1 \\ u_2 \end{bmatrix} + \begin{bmatrix} b_1 & b_2 \end{bmatrix} \begin{bmatrix} u_1 \\ u_2 \end{bmatrix} + a, \tag{2.33}$$

where q_1, q_2, q_3, b_1, b_2, and a depend on the electrical parameters and the state feedback gain matrix K. Moreover, as shown in Lemma 1 of [106], this quadratic function is always *strongly convex* because its associated Hessian matrix $\begin{bmatrix} q_1 & q_2 \\ q_2 & q_3 \end{bmatrix}$ is always *positive definite*. Since the mechanical dynamics (2.25), in ω_r, are driven by T_e, while T_e in (2.33) is a quadratic function of u_1 and u_2, the *two* new control variables u_1 and u_2 collectively affect *one* state variable ω_r. This implies that there is a redundancy in u_1 and u_2, which may be exploited elsewhere. Since the quadratic function is always strongly convex, this redundancy may be exposed via the following coordinate change [106], which transforms $u_1, u_2 \in \mathbb{R}$ in a Cartesian coordinate system into $r \geq 0$ and $\theta \in [-\pi, \pi)$ in a polar coordinate system

$$r = \sqrt{z_1^2 + z_2^2}, \quad \theta = \operatorname{atan2}(z_2, z_1), \tag{2.34}$$

where

$$\begin{bmatrix} z_1 \\ z_2 \end{bmatrix} = D^{1/2} M^T \begin{bmatrix} u_1 \\ u_2 \end{bmatrix} + \frac{1}{2} D^{-1/2} M^T \begin{bmatrix} b_1 \\ b_2 \end{bmatrix}, \tag{2.35}$$

atan2(·, ·) denotes the four-quadrant arctangent function, and M and D contain the eigenvectors and eigenvalues of $\left[\begin{smallmatrix} q_1 & q_2 \\ q_2 & q_3 \end{smallmatrix}\right]$ on their columns and diagonal, respectively, i.e.,

$$M^T \begin{bmatrix} q_1 & q_2 \\ q_2 & q_3 \end{bmatrix} M = D.$$

In the polar coordinates, it follows from (2.33)–(2.35) that

$$T_e = r^2 + a', \tag{2.36}$$

where $a' = -(v_{ds}^2 + v_{qs}^2)/(4\omega_s R_s)$ is always negative. From (2.25) and (2.36), we see that in the polar coordinates, r^2 is responsible for driving the mechanical dynamics in ω_r and, hence, may be viewed as an *equivalent electromagnetic torque*, differed from T_e only by a constant a'. On the other hand, the *polar angle* θ has no impact on the mechanical dynamics and, thus, represents the redundancy that will be exploited later, in Sect. 2.4.3.

Note that (2.34) and (2.35) describe the *Cartesian-to-Polar Coordinate Change* block in Fig. 2.2.

Having introduced the coordinate change, we next show that the unknown aerodynamic and mechanical parameters can be lumped into a scalar term, simplifying the problem. Combining (2.25), (2.27), (2.28), and (2.36),

$$J\dot{\omega}_r = \frac{\frac{1}{2}\rho A C_p(\frac{\omega_r R}{V_w}, \beta) V_w^3}{\omega_r} - r^2 - a' - C_f \omega_r. \tag{2.37}$$

Notice that the unknown parameters—namely, the C_p-surface, the air density ρ, the friction coefficient C_f, and the wind speed V_w—all appear in (2.37). Moreover, these unknown parameters can be separated from the "control input" r^2 and lumped into a scalar function $g(\omega_r, \beta, V_w)$, defined as

$$g(\omega_r, \beta, V_w) = \frac{\frac{1}{2}\rho A C_p(\frac{\omega_r R}{V_w}, \beta) V_w^3}{\omega_r} - a' - C_f \omega_r. \tag{2.38}$$

With $g(\omega_r, \beta, V_w)$ in (2.38) representing the aggregated uncertainties, the first-order dynamics (2.37) simplify to

$$\dot{\omega}_r = \frac{1}{J}(g(\omega_r, \beta, V_w) - r^2). \tag{2.39}$$

To design a controller for r^2, which allows the rotor angular velocity ω_r to track a desired, slow-varying reference ω_{rd} despite the unknown scalar function $g(\omega_r, \beta, V_w)$, consider a first-order nonlinear system

$$\dot{x} = \frac{1}{J}(f(x) + u), \tag{2.40}$$

where $x \in \mathbb{R}$ is the state, $u \in \mathbb{R}$ is the input, and $f(x)$ is a *known function* of x. Obviously, to drive x to some desired value $x_d \in \mathbb{R}$, we may apply feedback linearization [62] to cancel $f(x)$ and insert linear dynamics, i.e., let

$$u = -f(x) - \alpha(x - x_d), \tag{2.41}$$

where $\alpha \in \mathbb{R}$ is the controller gain. Combining (2.40) with (2.41) yields the closed-loop dynamics

$$\dot{x} = -\frac{\alpha}{J}(x - x_d). \tag{2.42}$$

Thus, if α is positive, x in (2.42) asymptotically goes to x_d.

Now suppose $f(x)$ in (2.40) is *unknown* but a *constant*, denoted simply as $f \in \mathbb{R}$ (we will relax the assumption that it is a constant shortly). With f being unknown, the controller (2.41) is no longer applicable. To overcome this limitation, we may first introduce a reduced-order estimator [17], which calculates an estimate $\widehat{f} \in \mathbb{R}$ of f, and then replace $f(x)$ in (2.41) by the estimate $\widehat{f}$

$$\dot{z} = -\frac{h}{J}(u + \widehat{f}), \tag{2.43}$$

$$\widehat{f} = z + hx, \tag{2.44}$$

$$u = -\widehat{f} - \alpha(x - x_d), \tag{2.45}$$

where $z \in \mathbb{R}$ is the estimator state and $h \in \mathbb{R}$ is the estimator gain. Defining the estimation error as $\widetilde{f} = f - \widehat{f}$ and combining (2.40) with (2.43)–(2.45) yield closed-loop dynamics

$$\dot{\tilde{f}} = -\dot{\widehat{f}} = -\dot{z} - h\dot{x} = -\frac{h}{J}\tilde{f}, \tag{2.46}$$

$$\dot{x} = \frac{1}{J}(f - \widehat{f} - \alpha(x - x_d)) = \frac{1}{J}(\tilde{f} - \alpha(x - x_d)). \tag{2.47}$$

Hence, by letting both α and h be positive, both $\tilde{f}$ and x in (2.46) and (2.47) asymptotically go to 0 and x_d, respectively.

Next, suppose both the state x and the desired value x_d must be *positive*, instead of being anywhere in $\mathbb{R}$. With this restriction, the controller with uncertainy estimation (2.43)–(2.45) needs to be modified, because for some initial conditions, it is possible that x can become nonpositive. One way to modify the controller is to replace the linear term $x - x_d$ in (2.45) by a logarithmic one $\ln(x/x_d)$, resulting in

$$u = -\widehat{f} - \alpha \ln \frac{x}{x_d}. \tag{2.48}$$

With (2.43), (2.44), and (2.48), the closed-loop dynamics become

$$\dot{\tilde{f}} = -\frac{h}{J}\tilde{f}, \tag{2.49}$$

$$\dot{x} = \frac{1}{J}(\tilde{f} - \alpha \ln \frac{x}{x_d}). \tag{2.50}$$

Note from (2.50) that for any $\tilde{f} \in \mathbb{R}$, there exists positive x, sufficiently small, such that $\dot{x}$ is positive. Therefore, for any initial condition $(\tilde{f}(0), x(0))$ with positive $x(0)$, $x(t)$ will remain positive, suggesting that the modification (2.48) satisfies the restriction that both x and x_d must be positive.

Now suppose the input u must be *nonpositive*. With this additional restriction, (2.48) needs to be further modified. One way to do so is to force the right-hand side of (2.48) to be nonpositive, leading to

$$u = -\max\left\{\widehat{f} + \alpha \ln \frac{x}{x_d}, 0\right\}. \tag{2.51}$$

Clearly, with (2.51), u is always nonpositive.

Finally, suppose f is an *unknown function* of x, denoted as $f(x)$. With this relaxation, we may associate the first-order nonlinear system (2.40) with the first-order dynamics (2.39) by viewing x as ω_r, x_d as ω_{rd}, u as $-r^2$, $f(x)$ as $g(\omega_r, \beta, V_w)$ (treating β and V_w as constants), and $\widehat{f}$ as $\widehat{g}$ (i.e., $\widehat{g}$ is an estimate of $g(\omega_r, \beta, V_w)$). Based on this association, (2.43), (2.44), and (2.51) can be written as

$$\dot{z} = -\frac{h}{J}(-r^2 + \widehat{g}), \tag{2.52}$$

$$\widehat{g} = z + h\omega_r, \tag{2.53}$$

$$r^2 = \max\left\{\widehat{g} + \alpha \ln \frac{\omega_r}{\omega_{rd}}, 0\right\}. \tag{2.54}$$

Having derived the controller with uncertainty estimation (2.52)–(2.54), we now analyze its behavior. To do so, some setup is needed: first, suppose ω_{rd}, β, and V_w are constants. Second, as shown in Lemma 2 of [106], because of Assumptions A1–A3 in Sect. 2.2, there exists $\omega_r^{(1)} \in (0, \infty)$ such that $g(\omega_r^{(1)}, \beta, V_w) = 0$ and $g(\omega_r, \beta, V_w) > 0$ for all $\omega_r \in (0, \omega_r^{(1)})$. Third, using (2.28), (2.38), and Assumptions A1 and A4 in Sect. 2.2, it is straightforward to show that there exist $\underline{\gamma} \in (-\infty, 0)$ and $\overline{\gamma} \in (0, \infty)$ such that $\underline{\gamma} \le \frac{\partial}{\partial \omega_r} g(\omega_r, \beta, V_w) \le \overline{\gamma}$ for all $\omega_r \in (0, \infty)$. Finally, with (2.39) and (2.52)–(2.54) and with $(\omega_r, \widehat{g})$ as state variables (instead of (ω_r, z)), the closed-loop dynamics can be expressed as

$$\dot{\omega}_r = \frac{1}{J}\left(g(\omega_r, \beta, V_w) - \max\left\{\widehat{g} + \alpha \ln \frac{\omega_r}{\omega_{rd}}, 0\right\}\right), \tag{2.55}$$

$$\dot{\widehat{g}} = \dot{z} + h\dot{\omega}_r = \frac{h}{J}(g(\omega_r, \beta, V_w) - \widehat{g}). \tag{2.56}$$

The following theorem characterizes the stability properties of the closed-loop system (2.55) and (2.56), the proof of which can be found in [48]:

Theorem 2.1 *Consider the closed-loop system (2.55) and (2.56). Suppose ω_{rd}, β, and V_w are constants with $0 < \omega_{rd} \le \omega_r^{(1)}$, where $\omega_r^{(1)}$, along with $\underline{\gamma}$ and $\overline{\gamma}$, is as defined above. Let $D = \{(\omega_r, \widehat{g}) | 0 < \omega_r \le \omega_r^{(1)}, \widehat{g} \in \mathbb{R}\} \subset \mathbb{R}^2$. If the controller gain α is positive and estimator gain h is large enough, i.e.,*

$$h > \overline{\gamma} \quad \textit{if} \quad \overline{\gamma} \ge -\frac{1}{3}\underline{\gamma}, \tag{2.57}$$

$$h > -\frac{(\overline{\gamma} - \underline{\gamma})^2}{8(\underline{\gamma} + \overline{\gamma})} \quad \textit{otherwise},$$

then: (i) the system has a unique equilibrium point at $(\omega_{rd}, g(\omega_{rd}, \beta, V_w))$ in D; (ii) the set D is a positively invariant set, i.e., if $(\omega_r(0), \widehat{g}(0)) \in D$, then $(\omega_r(t), \widehat{g}(t)) \in D$ $\forall t \ge 0$; and (iii) the equilibrium point $(\omega_{rd}, g(\omega_{rd}, \beta, V_w))$ is locally asymptotically stable with a domain of attraction D.

Theorem 2.1 says that, by using the electromagnetic torque subcontroller with uncertainty estimation (2.52)–(2.54), if the gains α and h are positive and sufficiently large and if the desired reference ω_{rd} does not exceed $\omega_r^{(1)}$, then the rotor angular velocity ω_r asymptotically converges to ω_{rd} if ω_{rd}, β, and V_w are constants and closely tracks ω_{rd} if they are slow-varying. Notice that the gains α and h can be chosen independently of each other. Also, the condition "$\omega_{rd} \le \omega_r^{(1)}$" is practically always satisfied, as $\omega_r^{(1)}$ is extremely large (see Fig. 3 of [106]).

Note that (2.52)–(2.54) describe the *Electromagnetic Torque Subcontroller with Uncertainty Estimation* block in Fig. 2.2.

2.4.3 Polar Angle and Desired Rotor Angular Velocity Subcontroller

Up to this point in the chapter, we have yet to specify how θ, ω_{rd}, and β are determined. To do so, we first introduce a scalar performance measure and express this measure as a function of θ, ω_{rd}, and β. We then present a method for choosing these variables, which optimizes the measure.

Recall that the ultimate goal is to make the active and reactive powers P and Q track some desired P_d and Q_d as closely as possible. Hence, it is useful to introduce a scalar performance measure, which characterizes how far P and Q are from P_d and Q_d. One such measure, denoted as U, is given by

$$U = \frac{1}{2}\begin{bmatrix} P - P_d & Q - Q_d \end{bmatrix}\begin{bmatrix} w_p & w_{pq} \\ w_{pq} & w_q \end{bmatrix}\begin{bmatrix} P - P_d \\ Q - Q_d \end{bmatrix}, \tag{2.58}$$

where w_p, w_q, and w_{pq} are design parameters satisfying $w_p > 0$ and $w_p w_q > w_{pq}^2$, so that $\left[\begin{smallmatrix} w_p & w_{pq} \\ w_{pq} & w_q \end{smallmatrix}\right]$ is a positive definite matrix. With these design parameters, one may specify how the differences $P - P_d$ and $Q - Q_d$ and their product $(P - P_d)(Q - Q_d)$ are penalized. Moreover, with U in (2.58) being a quadratic, positive definite function of $P - P_d$ and $Q - Q_d$, the smaller U is, the better the ultimate goal is achieved.

Having defined the performance measure U via (2.58), we next establish the following statement: if the subcontrollers in Sects. 2.4.1 and 2.4.2 are used with K chosen so that $A - BK$ has very fast eigenvalues, α chosen to be positive, and h chosen to satisfy (2.57), and if θ, ω_{rd}, β, V_w, P_d, and Q_d are all constants, then after a short transient, U may be expressed as a *known* function f_1 of r^2, θ, ω_{rd}, P_d, and Q_d, while r^2, in turn, may be expressed as an *unknown* function f_2 of ω_{rd}, β, and V_w, i.e.,

$$U = f_1(r^2, \theta, \omega_{rd}, P_d, Q_d), \tag{2.59}$$

$$r^2 = f_2(\omega_{rd}, \beta, V_w), \tag{2.60}$$

as shown in Fig. 2.3. To establish this statement, suppose the hypothesis is true. Then, after a short transient, it follows from (2.58) that U is a known function of P, Q, P_d, and Q_d; from (2.18)–(2.24), (2.29), and (2.30) that P and Q are known functions of φ, ω_r, u_1, and u_2; from (2.32) that φ is a known function of u_1 and u_2; from (2.34)

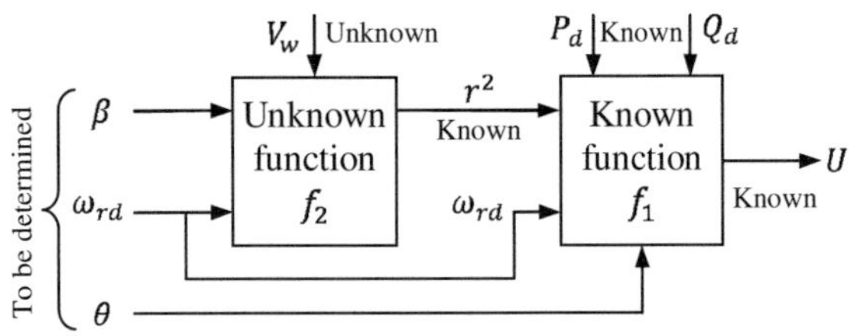

Fig. 2.3 Relationships among the performance measure U, the to-be-determined variables θ, ω_{rd}, and β, and the exogenous variables V_w, P_d, and Q_d

and (2.35) that u_1 and u_2 are known functions of r^2 and θ; and from Theorem 2.1 that $\omega_r = \omega_{rd}$. Thus, (2.59) holds with f_1 being known. Also, it follows from (2.54) and Theorem 2.1 that $r^2 = g(\omega_{rd}, \beta, V_w)$. Hence, (2.60) holds with f_2 being unknown.

Equations (2.59) and (2.60), which are represented in Fig. 2.3, suggest that U is a function of the to-be-determined variables θ, ω_{rd}, and β as well as the exogenous variables V_w, P_d, and Q_d. Given that the smaller U is the better, these to-be-determined variables may be chosen to minimize U. However, such minimization is difficult to carry out because although P_d and Q_d are known, V_w is not. To make matter worse, since f_1 is known but f_2 is not, the objective function is not entirely known. Somewhat fortunately, as shown in Fig. 2.3, θ affects U only through f_1 and not f_2. Therefore, θ may be chosen to minimize U for any given r^2, ω_{rd}, P_d, and Q_d, i.e.,

$$\theta = \arg\min_{x\in[-\pi,\pi)} f_1(r^2, x, \omega_{rd}, P_d, Q_d), \tag{2.61}$$

which is implementable since r^2, ω_{rd}, P_d, and Q_d are all known. Alternatively, θ may be chosen as in (2.61) but with a low-pass filter inserted to reduce possible chattering, such as a moving average filter, i.e.,

$$\theta(t) = \frac{1}{T_{ma}} \int_{t-T_{ma}}^{t} \arg\min_{x\in[-\pi,\pi)} f_1(r^2(\tau), x, \omega_{rd}(\tau), P_d(\tau), Q_d(\tau))d\tau, \tag{2.62}$$

where $T_{ma} > 0$ is the moving-average window size. With θ chosen as in (2.61), the minimization problem reduces from a three-dimensional problem to a two-dimensional one, depending only on ω_{rd} and β. Since the objective function upon absorbing θ is unknown and since V_w may change quickly, instead of minimizing U with respect to both ω_{rd} and β—which may take a long time—we decide to sacrifice freedom for speed, minimizing U only with respect to ω_{rd} and updating β in a relatively slower fashion, which will be described in Sect. 2.4.4.

The minimization of U with respect to ω_{rd} is carried out based on a gradient-like approach as shown in Fig. 2.4. To explain the rationale behind this approach, suppose β, V_w, P_d, and Q_d are constants. Then, according to (2.59)–(2.61), U is an unknown function of ω_{rd}. Because this function is not known, its gradient $\frac{\partial U}{\partial \omega_{rd}}$ at any ω_{rd} cannot be evaluated. To alleviate this issue, we evaluate U at two nearby ω_{rd}'s, use the two evaluated U's to obtain an estimate of the gradient $\frac{\partial U}{\partial \omega_{rd}}$, and move ω_{rd} along the direction where U decreases, by an amount which depends on the gradient estimate. This idea is illustrated in Fig. 2.4 and described precisely as follows: the desired rotor angular velocity $\omega_{rd}(t)$ is set to an initial value $\omega_{rd}(0)$ at time $t = 0$ and held constant until $t = T_1$, where T_1 should be sufficiently large so that both the electrical and mechanical dynamics have a chance to reach steady-state, but not too large which causes the minimization to be too slow. From time $t = T_1 - T_0$ to $t = T_1$, the average of $U(t)$, i.e., $\frac{1}{T_0}\int_{T_1-T_0}^{T_1} U(t)dt$, is recorded as the first value needed to obtain a gradient estimate. Similar to T_1, T_0 should be large enough so that small

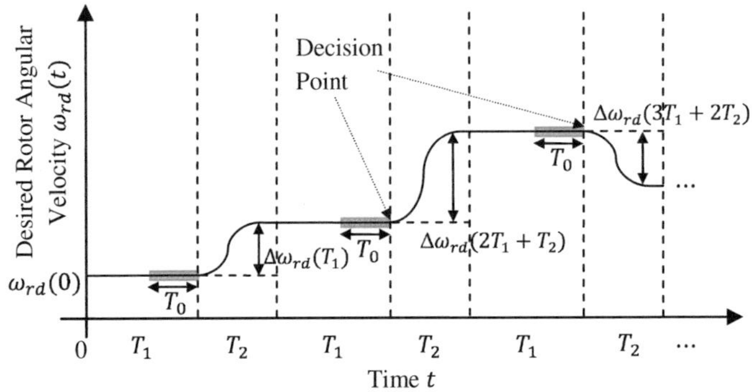

Fig. 2.4 Graphical illustration of the gradient-like approach

fluctuations in $U(t)$ (induced perhaps by a noisy V_w) are averaged out, but not too large which causes transient in the dynamics to be included. The variable $\omega_{rd}(t)$ is then changed gradually in an S-shape manner from $\omega_{rd}(0)$ at time $t = T_1$ to a nearby $\omega_{rd}(0)+\Delta\omega_{rd}(T_1)$ at $t = T_1+T_2$, where $\Delta\omega_{rd}(T_1)$ is an initial stepsize, and T_2 should be sufficiently large but not overly so, so that the transition in $\omega_{rd}(t)$ is smooth and yet not too slow. The variable $\omega_{rd}(t)$ is then held constant until $t = 2T_1 + T_2$, and the average of $U(t)$ from $t = 2T_1 + T_2 - T_0$ to $t = 2T_1 + T_2$, i.e., $\frac{1}{T_0}\int_{2T_1+T_2-T_0}^{2T_1+T_2} U(t)dt$, is recorded as the second value needed to obtain the gradient estimate. At time $t = 2T_1 + T_2$, the two recorded values are used to form the gradient estimate, which is in turn used to decide a new stepsize $\Delta\omega_{rd}(2T_1 + T_2)$ through

$$\Delta\omega_{rd}(2T_1 + T_2) = -\epsilon_1 \,\mathrm{sat}\left(\frac{\frac{1}{T_0}\int_{2T_1+T_2-T_0}^{2T_1+T_2} U(t)dt - \frac{1}{T_0}\int_{T_1-T_0}^{T_1} U(t)dt}{\epsilon_2 \Delta\omega_{rd}(T_1)}\right), \tag{2.63}$$

where $\epsilon_1 > 0$ and $\epsilon_2 > 0$ are design parameters that define the new stepsize $\Delta\omega_{rd}(2T_1 + T_2)$, and sat$(\cdot)$ is the standard saturation function that limits $\Delta\omega_{rd}(2T_1 + T_2)$ to $\pm\epsilon_1$. Upon deciding $\Delta\omega_{rd}(2T_1 + T_2)$, $\omega_{rd}(t)$ is again changed in an S-shape manner from $\omega_{rd}(0) + \Delta\omega_{rd}(T_1)$ at $t = 2T_1 + T_2$ to $\omega_{rd}(0) + \Delta\omega_{rd}(T_1) + \Delta\omega_{rd}(2T_1 + T_2)$ at $t = 2T_1 + 2T_2$, in a way similar to the time interval $[T_1, T_1 + T_2]$. The process then repeats with the second recorded value from the previous cycle $[0, 2T_1 + T_2]$ becoming the first recorded value for the next cycle $[T_1 + T_2, 3T_1 + 2T_2]$, and so on. Therefore, with this gradient-like approach, ω_{rd} is guaranteed to approach a local minimum when β, V_w, P_d, and Q_d are constants, and track a local minimum when they are slow-varying.

Note that (2.58), (2.61), and (2.63) describe the *Polar Angle and Desired Rotor Angular Velocity Subcontroller* block in Fig. 2.2.

2.4.4 *Blade Pitch Angle Subcontroller*

As mentioned, in order to speed up the minimization, we have decided to minimize U only with respect to ω_{rd}, leaving the blade pitch angle β as the remaining undetermined variable. Given that an active power P that is larger than the rated value P_{rated} of the turbine may cause damage, we decide to use β to prevent P from exceeding P_{rated}, thereby protecting the turbine. Specifically, we let β be updated according to

$$\dot{\beta} = \begin{cases} 0, & \text{if } \beta = \beta_{\min} \text{ and } P < P_{\text{rated}}, \\ 0, & \text{if } \beta = \beta_{\max} \text{ and } P > P_{\text{rated}}, \\ -\epsilon_3(P_{\text{rated}} - P), & \text{otherwise}, \end{cases} \tag{2.64}$$

where $\epsilon_3 > 0$ is a design parameter that dictates the rate at which β changes. Note that with (2.64), β is guaranteed to lie in $[\beta_{\min}, \beta_{\max}]$. Moreover, when P is above (below) P_{rated}, β increases (decreases) if possible, in order to try to capture less (more) wind power, which leads to a smaller (larger) P.

Note that (2.64) describes the *Blade Pitch Angle Subcontroller* block in Fig. 2.2. Also note that the blade pitch angle subcontroller may be designed based on other considerations. For example, if the forecast of, say, the hourly average wind speed $\overline{V}_w$ is available, for blade protection β may be chosen as $\beta = F(\overline{V}_w)$ for some nondecreasing function $F : (0, \infty) \to [\beta_{\min}, \beta_{\max}]$.

2.5 Simulation Results

In this section, we demonstrate the effectiveness of the proposed controller through simulation in MATLAB using a realistic wind profile from a wind farm in Oklahoma. To describe the simulation settings and results, both the per-unit and physical unit systems will be used interchangeably.

The simulation settings are as follows: we consider the General Electric (GE) 1.5 MW turbine adopted by the Distributed Resources Library in MATLAB/Simulink, which has a base voltage of 575 V and a base frequency of 60 Hz. The values of the turbine parameters are: $\omega_s = 1$ pu, $R_s = 0.00706$ pu, $R_r = 0.005$ pu, $L_s = 3.071$ pu, $L_r = 3.056$ pu, $L_m = 2.9$ pu, $v_{ds} = 1$ pu, $v_{qs} = 0$ pu, $J = 10.08$ pu, $A = 4656.6\,\text{m}^2$, $R = 38.5$ m, $\beta_{\min} = 0°$, $\beta_{\max} = 30°$, and $C_f = 0.01$ pu. The C_p-surface adopted by MATLAB, which is taken from [51], is $C_p(\lambda, \beta) = c_1\left(\frac{c_2}{\lambda_i} - c_3\beta - c_4\right)e^{\frac{-c_5}{\lambda_i}} + c_6\lambda$, where $\frac{1}{\lambda_i} = \frac{1}{\lambda+0.08\beta} - \frac{0.035}{\beta^3+1}$, $c_1 = 0.5176$, $c_2 = 116$, $c_3 = 0.4$, $c_4 = 5$, $c_5 = 21$, and $c_6 = 0.0068$. The mechanical power captured by the wind turbine is $P_m(\text{pu}) = \frac{P_{\text{nom}} P_{\text{wind\_base}}}{P_{\text{elec\_base}}} C_p(\text{pu}) V_w(\text{pu})^3$, where $P_m(\text{pu}) = \frac{P_m}{P_{\text{nom}}}$, $P_{\text{nom}} = 1.5$ MW is the nominal mechanical power, $P_{\text{wind_base}} = 0.73$pu is the maximum power at the base wind speed, $P_{\text{elec\_base}} = 1.5 \times 10^6/0.9$ VA is the base power of the electrical generator, $C_p(\text{pu}) = \frac{C_p}{C_{p\_\text{nom}}}$, $C_{p\_\text{nom}} = 0.48$ is the peak of the C_p-surface,

$V_w(\text{pu}) = \frac{V_w}{V_{w\_\text{base}}}$, and $V_{w\_\text{base}} = 12$ m/s is the base wind speed. Note that the maximum mechanical power, captured at the base wind speed, is 0.657 pu. The tip speed ratio is $\lambda(\text{pu}) = \frac{\frac{\omega_r(\text{pu})}{\omega_{r\_\text{base}}}}{V_w(\text{pu})}$, where $\lambda(\text{pu}) = \frac{\lambda}{\lambda_{\text{nom}}}$, $\lambda_{\text{nom}} = 8.1$ is the λ that yields the peak of the C_p-surface, $\omega_{r\_\text{base}} = 1.2$ pu is the base rotational speed, $\omega_r(\text{pu}) = \frac{\omega_r}{\omega_{r\_\text{nom}}}$, and $\omega_{r\_\text{nom}} = 2.1039$ rad/s is the nominal rotor angular velocity. For more details on these parameters and values, see the MATLAB documentation.

As for the proposed controller, we choose its parameters as follows: for the Rotor Voltages Subcontroller, we let the desired closed-loop eigenvalues of the electrical dynamics be at -5, $-10 \pm 5j$, and -15. Using MATLAB's place() function, the state feedback gain matrix $K = [K_1\ K_2]^T$ that yields these eigenvalues is found to be

$$K = \begin{bmatrix} 5135.9 & 259.2 & 20.3 & 1.9 \\ -2676.7 & 4289.9 & -1.3 & 19.7 \end{bmatrix}.$$

Moreover, we let $\alpha = 5$ and $h = 17.5$ for the Electromagnetic Torque Subcontroller with Uncertainty Estimation; let $w_p = 10$, $w_q = 1$, $w_{pq} = 0$, $\epsilon_1 = 0.025$, $\epsilon_2 = 2$, $T_0 = 1$ s, $T_1 = 4$ s, and $T_2 = 6$ s and use (2.62) with $T_{ma} = 0.75$ s for the Polar Angle and Desired Rotor Angular Velocity Subcontroller; and let $\epsilon_3 = 3$ and $P_{\text{rated}} = 1$ pu for the Blade Pitch Angle Subcontroller.

The simulation results are as follows: we consider a scenario where the wind speed V_w is derived from actual wind profiles from a wind farm located in northwest Oklahoma, the desired active power P_d experiences large step changes, and the desired reactive power Q_d is such that the desired power factor PF_d is fixed at 0.995. As will be explained below, these values of P_d force the turbine to operate in both the MPT and PR modes and switch between them, under realistic wind profiles. Figures 2.5 and 2.6 show the simulation results for this scenario in both the per-unit and physical unit systems, wherever applicable. Note that in Fig. 2.5, for the first 1200 s during which P_d is unachievable at 1 pu, the turbine operates in the MPT mode and maximizes P, as indicated by the value of C_p approaching its maximum of 0.48 after a short transient (the turbine is initially at rest). At time 1200 s when P_d drops sharply from 1 pu to an achievable value of 0.35 pu, the turbine quickly reduces the value of C_p, accurately regulates P around P_d, and effectively rejects the "disturbance" V_w, thereby smoothly switches from the MPT mode to the PR mode. At time 2400 s when P_d goes from 0.35 pu back to 1 pu, the MPT mode resumes. Because V_w is strong enough at that time, P approaches P_d. Moreover, the moment P exceeds P_d (which is equal to P_{rated}), the blade pitch angle β increases in order to clip the power and protect the turbine. At time 2700 s when V_w becomes weaker, β returns to $\beta_{\min} = 0°$, thereby allowing the value of C_p to return to its maximum of 0.48 and P to be maximized. As can be seen from the figure, throughout the simulation, PF is maintained near PF_d, affected only slightly and relatively shortly by the random wind fluctuations. Moreover, the angular velocity ω_r tracks the desired time-varying reference ω_{rd} closely. As expected, the small S-shape variations in ω_{rd} resemble those in Fig. 2.4. Notice that similar observations can be made in Fig. 2.6,

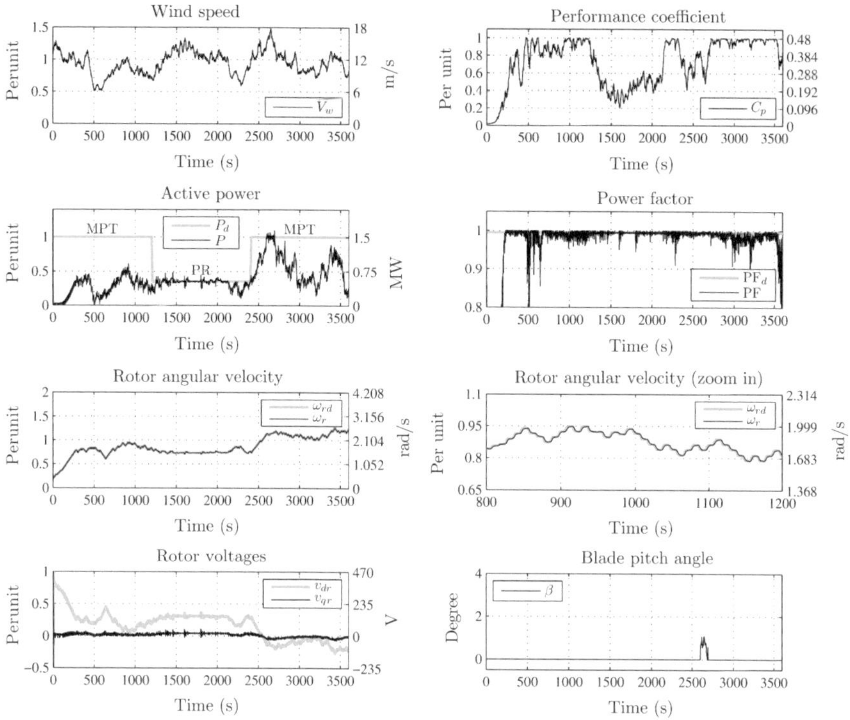

Fig. 2.5 Effective operation in both the MPT and PR modes and seamless switching between them under an actual wind profile from a wind farm located in northwest Oklahoma

which shows additional simulation results with a different wind profile and different desired active and reactive powers.

The above simulation results suggest that the proposed controller not only is capable of operating effectively in both the MPT and PR modes, it is also capable of switching smoothly between them—all the while not knowing the C_p-surface, air density, friction coefficient, and wind speed.

2.6 Concluding Remarks

In this chapter, we have described a mathematical model for variable-speed wind turbines employing DFIGs. Based on this model, we have utilized a number of control techniques—including feedback linearization, model order reduction, uncertainty estimation, and potential function minimization—to design a reconfigurable nonlinear dual-mode controller, which enables two most desirable operating modes of MPT and PR, while addressing other issues such as uncertainties in most of the

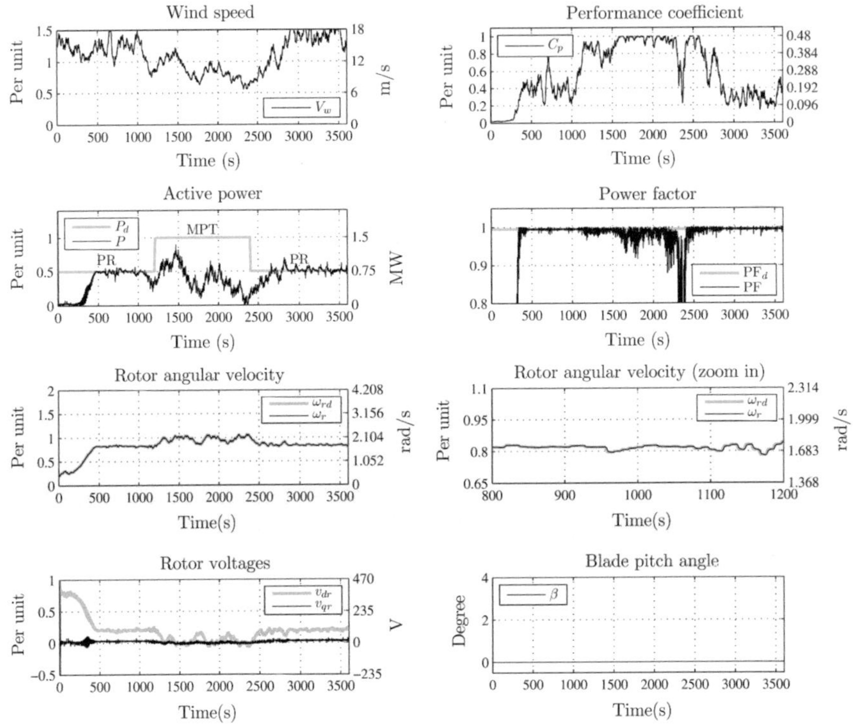

Fig. 2.6 Operation with a different wind profile and different desired active and reactive powers

model parameters. We have also demonstrated the effectiveness of this single-turbine controller through simulation with a realistic wind profile.

The work presented in this chapter, as well as those by many others, indicate that such reconfigurable controllers can expand the ability of active power control in meeting a number of rigid requirements on modern wind turbines. For example, when the above reconfigurable controller operates in the PR mode, the desired level of power output can be quickly reached and maintained, whenever wind is available. A controller equipped with such a PR mode allows the power produced from wind to become a source of secondary and tertiary controls for grid frequency support, or a source of up/down power balancing based on electricity market conditions. In addition, the PR mode offers turbines the flexibility to coordinate with other energy resources including storage and distributed generation.

Single-turbine control technologies have been at the center stage of research for many years. However, there are still a lot of challenges to overcome to fully meet the requirements of future power industry. For example, even for the well-known issue of using active power control to maintain the stability of grid frequency, it is not clear whether and to what extent wind farms can offer the full range of frequency support usually provided by conventional generation. Since electricity is not storable in large quantities, at any given point in time, the amount of electricity produced must be equal to the amount consumed to maintain a constant grid frequency. When there is a dramatic imbalance between electricity production and consumption, frequency deviations occur, which may cause serious stability problems or even cascading failures. Due to constantly changing loads and inevitable events such as forced outages and natural disasters, a certain amount of reserves and fast generation support must be procured to stabilize and/or restore the frequency. Frequency support includes inertia response as well as primary, secondary, and tertiary controls that require different amount of power and response times. Although the above reconfigurable controller and those by others can relieve some of the concerns on secondary and tertiary controls, it is unclear whether they can address concerns regarding other frequency support.

Nevertheless, active power control at the turbine level is not the focus of this monograph. Instead, the concepts and functions of active power control introduced in this chapter set the stage for discussion of active power control at the wind farm level in subsequent chapters. Specifically, we will address a number of essential questions at the wind farm level such as: how should wind turbines be modeled to facilitate the design and analysis of wind farm controllers? How does a wind farm controller make the turbines cooperate to achieve a better performance? What are the potential of wind farm control in enhancing wind power sustainability, and what are the open challenges in this regard?

Chapter 3
Approximate Model for Wind Farm Power Control

Abstract Wind farm power control is key to reliable large-scale wind integration. The design of a sophisticated wind farm controller, however, is nontrivial partly because there is a lack of models that appropriately simplify the complex overall dynamics of a large number of *wind turbine control systems* (WTCSs). In this chapter, we use system identification techniques to develop a simple *approximate model* that attempts to mimic the active and reactive power dynamics of two generic WTCS models under normal operating conditions: an analytical model described by nonlinear differential equations, and an empirical one by input–output measurement data. The approximate model contains two parts—one for active power and one for reactive power—each of which is a third-order system that would have been linear if not for a static nonlinearity. For each generic model, we also provide an identification scheme that sequentially determines the approximate model parameters. Finally, we show via simulation that, despite its structural simplicity, the approximate model is sufficient and versatile, capable of closely imitating several different analytical and empirical WTCS models from the literature and from real data. The results suggest that the approximate model may be used to facilitate research in wind farm power control, which we will carry out in subsequent chapters.

3.1 Introduction

Being able to control a wind farm so that its power output is cooperatively maximized, or smoothly regulated, is imperative to successful and reliable integration of large-scale wind generation into the power grid. The design of a sophisticated *wind farm control system* (WFCS) for such control, however, is challenging for a variety of reasons. First, as shown in Sect. 2.2, a wind turbine, by itself, is already a fairly complex system with highly nonlinear dynamics, strong electromechanical coupling, inherently uncertain parameters, and multiple control variables. Second, when hundreds of such turbines are immersed in a wind field across a geographical region, they produce turbulence and wake effects that affect downstream turbines, causing their overall behavior to be complicated. Third, the large number of control variables to simultaneously handle, and the rich set of approaches to possibly use,

J.N. Jiang et al., *Control and Operation of Grid-Connected Wind Farms*,
Advances in Industrial Control, DOI 10.1007/978-3-319-39135-9_3

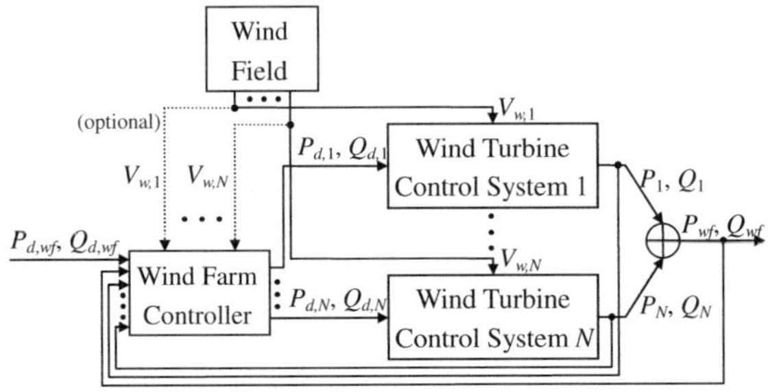

Fig. 3.1 Hierarchical architecture of a wind farm control system

further compound the complexity. Thus, it is challenging to design a WFCS. Indeed, there has been relatively little work to date on the topic [32, 40, 50, 70, 81, 93, 101, 103, 104, 107], compared to single-turbine control (e.g., [12, 23, 25, 31, 33, 42, 48, 49, 59, 69,73, 106] and many more).

One way to cope with the complexities, adopted in [32, 40, 50, 70, 81, 93, 101, 103, 104, 107], is to introduce a hierarchical architecture, which as shown in Fig. 3.1 divides a WFCS into two parts: a central *wind farm controller* (WFC) and N individual *wind turbine control systems* (WTCSs), each comprising a wind turbine and its controller. With this architecture, we may first design, for each $i \in \{1, 2, \ldots, N\}$, a WTCS i that tries to regulate its millisecond-to-second-timescale active and reactive power outputs P_i and Q_i at some desired $P_{d,i}$ and $Q_{d,i}$, regardless of its incoming wind speed $V_{w,i}$. Upon completion, we may then design a WFC that tries to regulate the second-to-minute-timescale wind farm power outputs $P_{wf} \triangleq \sum_{i=1}^{N} P_i$ and $Q_{wf} \triangleq \sum_{i=1}^{N} Q_i$ at some desired $P_{d,wf}$ and $Q_{d,wf}$, presumably from a grid operator, by adjusting the $P_{d,i}$'s and $Q_{d,i}$'s based on feedback of the P_i's and Q_i's and possibly estimates of the $V_{w,i}$'s. Hence, the architecture simplifies the design of a WFCS, allowing us to sequentially tackle two (seemingly) easier problems on different timescales, as opposed to tackling a harder one. The architecture also gives us the option of designing a new single-turbine controller for each WTCS i, or applying an existing one (e.g., [23, 25, 33, 48, 49, 73, 106]) that accepts $P_{d,i}$ and $Q_{d,i}$ as inputs.[1] Furthermore, it allows us to view the WFC as a second-to-minute-timescale supervisor that tells every WTCS i how much power to generate, and focus on its design without delving too much into millisecond-to-second, turbine-level details.

Although the hierarchical architecture makes the problem more manageable, it does not remove the fact that each WTCS—being a composition of an already-complex wind turbine and a possibly-complicated controller—typically has complex dynamics. As a result, the subsequent design and analysis of a supervisory WFC may prove to be difficult, depending on our goal: if we are content with a simple design (e.g., distribute $P_{d,wf}$ and $Q_{d,wf}$ evenly among the $P_{d,i}$'s and $Q_{d,i}$'s, or $P_{d,wf}$ proportionally based on the $V_{w,i}$'s) and a basic analysis (e.g., simulation studies only), then how complex a WTCS is probably does not matter. However, if we aim

[1] Single-turbine controllers that do not accept $P_{d,i}$ and $Q_{d,i}$, such as those that always attempt maximum power tracking (e.g., [12, 31, 42, 59, 69]), may not fit well with this architecture.

for a nifty design (e.g., adjust the $P_{d,i}$'s and $Q_{d,i}$'s so that the WTCSs can exploit their correlation, interaction, and/or diversity to cooperatively achieve faster transient responses and better steady-state smoothness in P_{wf} and Q_{wf}) and a deeper analysis and understanding (e.g., theoretical characterization of the resulting transient and steady-state behaviors), then an overly complex WTCS may render the process very difficult or even impossible. Therefore, to achieve the latter, it is necessary to build a suitably simplified WTCS model.

To this end, suppose we have developed, or are given, a WTCS—call it WTCS*—and wish to design a WFC. Also suppose, at our disposal, is a mathematical model parameterized by a vector θ—call it WTCS$^\theta$—which, like WTCS* (or each WTCS in Fig. 3.1), maps inputs (P_d, Q_d, V_w) to outputs (P, Q), i.e., $(P, Q) = \text{WTCS}^\theta(P_d, Q_d, V_w)$. Consider the following conditions on the model WTCS$^\theta$:

C1. There exists a θ such that whenever WTCS$^\theta$ and WTCS* are driven by the same inputs (P_d, Q_d, V_w), they produce approximately the same outputs (P, Q).
C2. WTCS$^\theta$ may be a nonlinear dynamical system but has a favorable structure conducive to control systems analysis and design.
C3. There is a set of WTCSs in the literature such that for each WTCS in the set, there exists a θ such that whenever WTCS$^\theta$ and the WTCS are driven by the same inputs (P_d, Q_d, V_w), they produce approximately the same outputs (P, Q).

Note that if Condition C1 holds, WTCS$^\theta$—with the specific value of θ—would be an accurate approximation of WTCS* and, thus, may be used in place of WTCS* in the WFC design and WFCS analysis. If, in addition, Condition C2 holds, the design and analysis would be more likely to succeed due to the favorable structure of WTCS$^\theta$. If Condition C3 holds as well, WTCS$^\theta$—with different values of θ—would be able to also approximate a number of different WTCSs in the literature (or by different manufacturers), making it a versatile model that brings WTCS* and those WTCSs under the same umbrella, distinguished only by θ. It follows that the design and analysis outcomes (e.g., new control techniques, stability criteria, and performance formulas) are applicable not only to WTCS*, but perhaps also to those WTCSs, increasing their impact. Hence, having an approximate model WTCS$^\theta$ that satisfies Conditions C1–C3 is very valuable.

This chapter is devoted to the development of such a model. We first assume, in Sect. 3.2, that two generic models of WTCS* are given, namely, an *analytical model* described by a set of continuous-time nonlinear differential equations, and an *empirical model* described by a set of input–output measurement data. The latter is motivated by the fact that in practice, what is available may just be a set of data, rather than a mathematical model, due to legacy and proprietary reasons. Based on standard system identification approaches [71] and typical WTCS characteristics, we then develop, in Sect. 3.3, an *approximate model* WTCS$^\theta$, which attempts to imitate both the analytical and empirical models. For each of these two models, we also provide a parameter identification scheme that sequentially determines the θ required by Condition C1, which turns out to be a vector of 10 parameters (two functions and eight scalars). The approximate model, depicted in Fig. 3.3c, may be regarded as satisfying Condition C2 because it is made up of two structurally identical parts—one for active power and the other for reactive—each of which is a third-order system that would have been linear if not for a static nonlinear component at its inputs (i.e., a modified Hammerstein model [71, 84]). Next, we validate, in Sect. 3.4, the approximate model via simulation, showing that it has enough ingredients to closely imitate several different analytical and empirical models from the literature [40, 59, 93, 106] and from real data taken from an Oklahoma wind farm. The encouraging results suggest that the approximate model satisfies Condition C3 and, hence, may be used to facilitate the development of a second-to-minute-timescale supervisory WFC that yields a sophisticated WFCS.

We stress that this chapter is not about *wind turbine* modeling (as in, say, [38, 66, 68]), controller design (as in, say, [12, 23, 25, 31, 33, 42, 48, 49, 59, 69, 73, 106]), and controller comparison [40], nor is it about *wind farm* modeling [35, 39, 41, 102] and controller design [32, 40, 50, 70, 72, 74, 81, 93, 96, 101, 103, 104, 107]. Rather, the chapter is on *wind turbine control system* modeling—not just the turbine, but the turbine *plus* its controller—in the context of the architecture in Fig. 3.1, for which there seems to be no prior work. We note that there has been significant efforts by the Western Electricity Coordinating Council (WECC), Utility Wind Integration Group (UWIG), and IEEE on developing generic WTCS models [3, 8, 99]. Such models, however, are intended for accurately simulating the impact of wind turbine generator dynamics on power system transient stability without compromising vendor proprietary information, rather than for facilitating research on wind farm power control in the aforementioned context, where model simplicity is at premium. Therefore, those models and the approximate model have notably different purposes. Indeed, they have very different structure and level of details, which can be seen by comparing, for instance, Figs. 1, 3, and 20–24 of [8] with Fig. 3.3c of this chapter.

3.2 Models of Wind Turbine Control Systems

In this section, we describe a generic analytical model and a generic empirical model of a WTCS under normal operating conditions. These two models set the stage for the development of an approximate model that mimics the dynamic performance of the WTCS, when it provides primary generation services to the grid.

3.2.1 Analytical Model

To control a variable-speed wind turbine, a standard approach is to first model its dynamics based on first principles, and then design a controller based on known techniques. Regardless of the model and design, the resulting WTCS typically can be represented in a block diagram form as in Fig. 3.2, and described generically by a set of continuous-time, nonlinear differential equations in state-space form as follows:

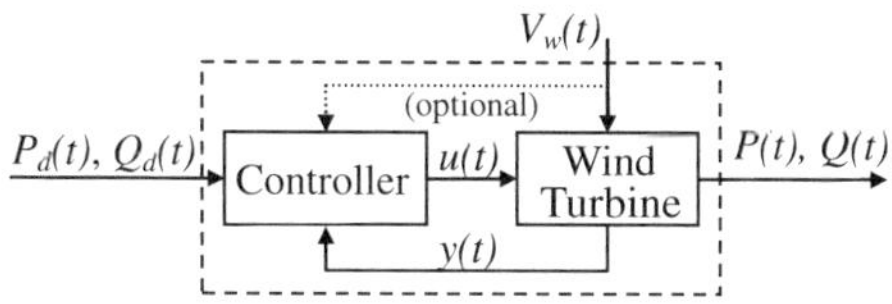

Fig. 3.2 Generic block diagram of a wind turbine control system

$$\dot{x}(t) = f(x(t), P_d(t), Q_d(t), V_w(t)), \quad x(0) = x_0, \tag{3.1}$$

$$(P(t), Q(t)) = g(x(t), P_d(t), Q_d(t), V_w(t)). \tag{3.2}$$

Here, $t \geq 0$ denotes time; $x(t) \in \mathbb{R}^n$ is the system states that combine the wind turbine states (e.g., stator and rotor fluxes or currents, rotor angular velocity) and controller states (if any); $x(0)$ is the initial states; f and g are functions depending on the particular wind turbine model (e.g., fourth-order [38] or second-order [68] doubly fed induction generator (DFIG) dynamics or second-order permanent magnet synchronous generator (PMSG) dynamics [66], and rigid-shaft or flexible-shaft [68] mechanical dynamics) and the particular controller design (e.g., one of the designs in [23, 25, 32, 33, 48–50, 73, 93, 106]) including their parameters (e.g., resistances, inductances, rotor moment of inertia, rotor swept area, C_p-surface, air density, friction coefficient, controller gains); $P_d(t)$, $Q_d(t)$, $P(t)$, and $Q(t)$ are, respectively, the system inputs and outputs representing the desired and actual active and reactive powers, where positive values mean toward the grid; $V_w(t)$ is another system input representing the wind speed; $u(t)$ is the internal control signals (e.g., rotor voltages, blade pitch angle, electromagnetic torque); and $y(t)$ is the internal feedback signals (e.g., various voltages and currents, rotor angular velocity, actual powers).

In this chapter, we assume that a generic *analytical model* of a WTCS, in the form of (3.1) and (3.2), is given as the first of two models considered. In order to represent as many WTCSs in the literature as possible, we make only two assumptions about the analytical model (3.1) and (3.2): first, the inputs $(P_d(t), Q_d(t), V_w(t))$ are always in a bounded *operating region*

$$\mathcal{S} \triangleq [0, P_{d,\mathrm{MAX}}] \times [-Q_{d,\mathrm{MAX}}, Q_{d,\mathrm{MAX}}] \times [0, V_{w,\mathrm{MAX}}].$$

Second, the WTCS is reasonably well-designed, i.e., the functions f and g are such that for each constant $(P_d(t), Q_d(t), V_w(t)) = (\overline{P}_d, \overline{Q}_d, \overline{V}_w) \in \mathcal{S}$, there exist steady-state values (P_{ss}, Q_{ss}), depending possibly on $(\overline{P}_d, \overline{Q}_d, \overline{V}_w)$, such that for all initial states $x(0) \in \mathbb{R}^n$,

$$\lim_{t\to\infty} (P(t), Q(t)) = (P_{ss}, Q_{ss}).$$

Finally, we allow $Q_d(t)$ and $Q(t)$ to be absent from (3.1) and (3.2), since many existing WTCSs do not consider the reactive power (e.g., [23, 25, 73]), and $P_d(t)$ to be absent as well, since some existing WTCSs do not require it to be specified (e.g.,

[12, 31, 42, 59, 69]). However, we require $P(t)$ and $V_w(t)$ to be present, since they are essential to WTCSs.

3.2.2 *Empirical Model*

Although it is common to work with a mathematical model in research on WTCSs, in practice we may not have access to the inner working of a WTCS, due perhaps to legacy and proprietary reasons. Instead, what may be available to us is a set of input–output measurement data, so that we have no choice but to treat the WTCS as a "black box." The set of data can take various forms, but more often than not includes the following information:

Inputs			Outputs	
$P_d(0)$	$Q_d(0)$	$V_w(0)$	$P(0)$	$Q(0)$
$P_d(\Delta)$	$Q_d(\Delta)$	$V_w(\Delta)$	$P(\Delta)$	$Q(\Delta)$
$\vdots$	$\vdots$	$\vdots$	$\vdots$	$\vdots$
$P_d((D-1)\Delta)$	$Q_d((D-1)\Delta)$	$V_w((D-1)\Delta)$	$P((D-1)\Delta)$	$Q((D-1)\Delta)$

(3.3)

where $\Delta > 0$ is the sampling period which is usually on the order of seconds or minutes, and D is the number of data points which is usually large.

In this chapter, we assume that a generic *empirical model* of a WTCS, in the form of (3.3), is given as the second of the two models considered. Similar to the analytical model (3.1) and (3.2), in the empirical model (3.3) the columns $P_d(i\Delta)$, $Q_d(i\Delta)$, and $Q(i\Delta)$ are optional but the columns $V_w(i\Delta)$ and $P(i\Delta)$ are mandatory. However, unlike the analytical one where the inputs $(P_d(t), Q_d(t), V_w(t))$ can be arbitrarily specified, with this empirical model we have no control over the inputs $(P_d(i\Delta), Q_d(i\Delta), V_w(i\Delta))$, as they are simply given, in the first three columns. This difference will be accounted for shortly.

Remark 3.1 The two models of WTCSs in this section may be thought of as the WTCS* in Sect. 3.1.

3.2.3 *Discussion*

The analytical model (3.1) and (3.2) and the empirical model (3.3) are what the approximate model intends to imitate. As will be described in Sect. 3.3, the approach is based on postulating a static nonlinear model that matches the steady-state input–output characteristics, followed by enriching the model with linear dynamics so that it also matches the transient input–output behaviors. A benefit of this input–output approach is that it bypasses the need to consider the internal dynamics and specific details of the underlying WTCS, thereby allowing major types of generation technologies such as DFIGs, PMSGs, and basic induction generators to be approximately described using a simple, consistent model. More importantly, such a model enables one to approximately describe a large number of same or different types of WTCSs within a wind farm in a unified fashion, so that researchers may focus on other pressing issues when designing a comprehensive WFCS and understanding its attainable performance.

3.3 Proposed Approximate Model

In this section, we develop a simple mathematical model that approximates the analytical and empirical WTCS models discussed in Sect. 3.2, and a parameter identification scheme that determines the model parameters in each case. The development consists of three steps in both cases, as described below.

3.3.1 *Approximating the Analytical Model*

Step 1: Mimicking the steady-state responses to constant inputs

In general, to create a system that mimics another system, it is reasonable to demand that the two systems exhibit the same steady-state responses to constant inputs. With this in mind, we note that whenever the analytical model (3.1) and (3.2) is subject to constant inputs $(P_d(t), Q_d(t), V_w(t)) = (\overline{P}_d, \overline{Q}_d, \overline{V}_w) \in \mathcal{S}$, its outputs $(P(t), Q(t))$ is assumed to asymptotically converge to some steady-state values (P_{ss}, Q_{ss}), which depend only on $(\overline{P}_d, \overline{Q}_d, \overline{V}_w)$ and not on the initial states $x(0)$. This dependency suggests that there exist functions $\varphi_1 : \mathcal{S} \to \mathbb{R}$ and $\varphi_2 : \mathcal{S} \to \mathbb{R}$, such that $P_{ss} = \varphi_1(\overline{P}_d, \overline{Q}_d, \overline{V}_w)$ and $Q_{ss} = \varphi_2(\overline{P}_d, \overline{Q}_d, \overline{V}_w)$. It also suggests a static nonlinear model of the form

$$P(t) = \varphi_1(P_d(t), Q_d(t), V_w(t)), \tag{3.4}$$

$$Q(t) = \varphi_2(P_d(t), Q_d(t), V_w(t)), \tag{3.5}$$

which is capable of mimicking—at the very least—the steady-state outputs of the analytical model (3.1) and (3.2) whenever the inputs are constant, or slow-varying. To visually connect this Step 1 with subsequent steps of the development, a block diagram of the model (3.4) and (3.5) is shown in Fig. 3.3a.

Remark 3.2 Throughout the chapter, the subscripts 1 and 2 are used to distinguish between similar parameters or variables associated with the *active* and *reactive* powers (e.g., φ_1 is for active and φ_2 is for reactive in Fig. 3.3a).

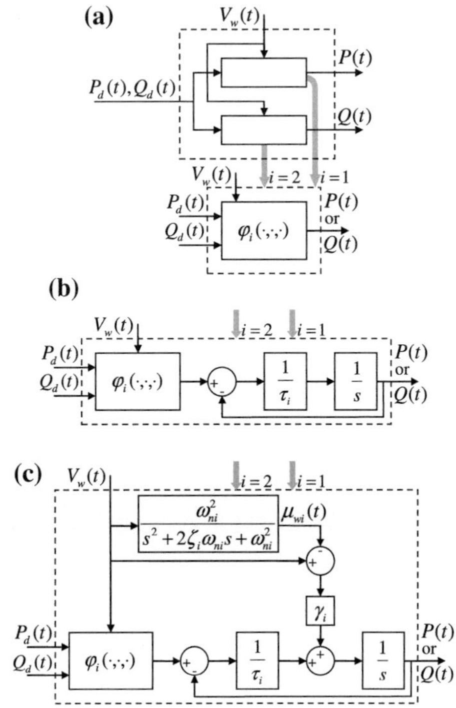

Fig. 3.3 Step-by-step development of the proposed approximate model. Block diagram after Step 1 of 3 (**a**). Block diagram after Step 2 of 3 (**b**). Block diagram after Step 3 of 3 (**c**)

The functions φ_1 and φ_2 are (infinite-dimensional) parameters of the model (3.4) and (3.5), which can be identified by simulating the analytical model (3.1) and (3.2) with various constant inputs sufficiently covering the operating region $\mathcal{S}$, observing the steady-state outputs, and employing interpolation. The following procedure provides the details:

Procedure for Step 1

---

(1) Pick a large $N_1 \in \mathbb{N} = \{1, 2, \ldots\}$.
(2) Pick $(P_d^{(i)}, Q_d^{(i)}, V_w^{(i)}) \in \mathcal{S}$ for $i = 1, 2, \ldots, N_1$.
(3) Pick a large $T > 0$.
(4) Loop over $i = 1, 2, \ldots, N_1$.
(5) Let $(P_d(t), Q_d(t), V_w(t)) = (P_d^{(i)}, Q_d^{(i)}, V_w^{(i)})\ \forall t \in [0, T]$.
(6) Pick any $x(0)$.
(7) Simulate the analytical model (3.1) and (3.2) from $t = 0$ to $t = T$.
(8) Record $(P(T), Q(T))$.
(9) Let $\varphi_1(P_d^{(i)}, Q_d^{(i)}, V_w^{(i)}) = P(T)$ and $\varphi_2(P_d^{(i)}, Q_d^{(i)}, V_w^{(i)}) = Q(T)$.
(10) End loop.
(11) Determine φ_1 and φ_2 via interpolation on the N_1 data points obtained.

Applying the above procedure to identify the functions φ_1 and φ_2, we obtain a basic model (3.4) and (3.5) that exhibits the same steady-state behavior as that of (3.1) and (3.2).

Step 2: Mimicking the transient responses to staircase inputs

The basic model (3.4) and (3.5) in Fig. 3.3a is able to match the steady-state response of the analytical model (3.1) and (3.2). However, it fails to produce any kind of transient one normally would expect with WTCSs because (3.4) and (3.5) are merely static functions mapping the inputs $(P_d(t), Q_d(t), V_w(t))$ to the outputs $(P(t), Q(t))$. To alleviate this drawback, we insert into Fig. 3.3a first-order linear dynamics between $\varphi_1(\cdot)$ and $P(t)$ and between $\varphi_2(\cdot)$ and $Q(t)$, to arrive at a Hammerstein model (see [84] and Chap. 5.2 of [71]) shown in Fig. 3.3b and given by

$$\dot{P}(t) = -\frac{1}{\tau_1}P(t) + \frac{1}{\tau_1}\varphi_1(P_d(t), Q_d(t), V_w(t)), \tag{3.6}$$

$$\dot{Q}(t) = -\frac{1}{\tau_2}Q(t) + \frac{1}{\tau_2}\varphi_2(P_d(t), Q_d(t), V_w(t)), \tag{3.7}$$

where $\tau_1 > 0$ and $\tau_2 > 0$ are the time constants. Note that in steady-state, (3.6) and (3.7) reduce to (3.4) and (3.5). Hence, (3.6) and (3.7) are able to capture not only the steady-state behavior of the analytical model (3.1) and (3.2), but also the dominant mode of its transient behavior with proper choices of τ_1 and τ_2.

The time constants τ_1 and τ_2 can be identified using a general approach in system identification sometimes known as the prediction-error methods (see Chap. 7 of [71]). With this approach, we first choose specific inputs and use them to simulate the analytical model (3.1) and (3.2) and the model (3.6) and (3.7), the latter with different values of τ_1 and τ_2. We then compare the outputs of the two models and determine the best τ_1 and τ_2, which minimize the output differences. The following procedure details this approach, in which we choose the inputs to be random staircase signals because they tend to bring out the dominant mode in systems, and allow any $\mathcal{L}_p$ norm to be used for measuring the output differences:

Procedure for Step 2

---

(1) Pick a large $N_2 \in \mathbb{N}$.
(2) Pick $(P_d^{(i)}, Q_d^{(i)}, V_w^{(i)}) \in \mathcal{S}$ for $i = 1, 2, \ldots, N_2$ randomly, independently, and equiprobably.
(3) Use the T in Step 1.
(4) Let $(P_d(t), Q_d(t), V_w(t)) = (P_d^{(i)}, Q_d^{(i)}, V_w^{(i)})$ $\forall t \in [(i-1)T, iT)$ for $i = 1, 2, \ldots, N_2$.
(5) Pick any $x(0)$.
(6) Simulate the analytical model (3.1) and (3.2) from $t = 0$ to $t = N_2 T$.
(7) Record $(P(t), Q(t))$ $\forall t \in [0, N_2 T]$ as $(P_{an}(t), Q_{an}(t))$.
(8) Pick a large $N_3 \in \mathbb{N}$.
(9) Pick $\tau_1^{(i)} > 0$ and $\tau_2^{(i)} > 0$ for $i = 1, 2, \ldots, N_3$.

(10) Use the φ_1 and φ_2 identified in Step 1.
(11) Pick any $(P(0), Q(0))$.
(12) Loop over $i = 1, 2, \ldots, N_3$.
(13) Let $\tau_1 = \tau_1^{(i)}$ and $\tau_2 = \tau_2^{(i)}$.
(14) Simulate the model (3.6) and (3.7) from $t = 0$ to $t = N_2 T$.
(15) Record $(P(t), Q(t))\ \forall t \in [0, N_2 T]$ as $(P_{ap}^{(i)}(t), Q_{ap}^{(i)}(t))$.
(16) Calculate

$$J_1(\tau_1^{(i)}) = \left(\int_{T_1}^{N_2 T} |P_{an}(t) - P_{ap}^{(i)}(t)|^{p_1} dt \right)^{\frac{1}{p_1}}$$

and

$$J_2(\tau_2^{(i)}) = \left(\int_{T_2}^{N_2 T} |Q_{an}(t) - Q_{ap}^{(i)}(t)|^{p_2} dt \right)^{\frac{1}{p_2}},$$

where $T_1 \in (0, N_2 T)$, $T_2 \in (0, N_2 T)$, $p_1 \geq 1$, and $p_2 \geq 1$.
(17) End loop.
(18) Let $\tau_1 = \underset{\tau \in \{\tau_1^{(1)}, \ldots, \tau_1^{(N_3)}\}}{\arg\min} J_1(\tau)$ and $\tau_2 = \underset{\tau \in \{\tau_2^{(1)}, \ldots, \tau_2^{(N_3)}\}}{\arg\min} J_2(\tau)$.

Remark 3.3 In the above procedure, $\tau_1^{(i)}$ and $\tau_2^{(i)}$ for $i = 1, 2, \ldots, N_3$ represent the search space for the best τ_1 and τ_2; p_1 and p_2 represent the desired $\mathcal{L}_p$ norms; and T_1 and T_2 are introduced to reduce the impact of the initial states (i.e., $x(0)$ of the analytical model (3.1) and (3.2) and $(P(0), Q(0))$ of the model (3.6) and (3.7)) on the parameter estimation process.

Using the preceding procedure to identify the time constants τ_1 and τ_2, we obtain a refined model (3.6) and (3.7) that has more flexibility to better match the behavior of (3.1) and (3.2).

Step 3: Mimicking the responses to realistic inputs

Although the refined model (3.6) and (3.7) in Fig. 3.3b is more sophisticated than the basic model (3.4) and (3.5) in Fig. 3.3a, it can only produce first-order-like responses. If such responses are indeed what the analytical model (3.1) and (3.2) produces, or if what we desire is just a crude approximation, then the refined model (3.6) and (3.7) may be satisfactory. Otherwise, its accuracy may be unacceptable.

At first glance, this issue can be overcome by replacing the first-order linear dynamics in (3.6) and (3.7) with higher-order ones. This approach, however, has a fundamental limitation: recall from Step 1 that $P_{ss} = \varphi_1(\overline{P}_d, \overline{Q}_d, \overline{V}_w)$ and $Q_{ss} =$

$\varphi_2(\overline{P}_d, \overline{Q}_d, \overline{V}_w)$. Thus, if a WTCS does power regulation and does it well over a wide range of $\overline{V}_w$, then $P_{ss} \approx \overline{P}_d$ and $Q_{ss} \approx \overline{Q}_d$ for any $\overline{V}_w$ in that range. As a result, $\varphi_1(P_d(t), Q_d(t), V_w(t))$ in (3.6) and $\varphi_2(P_d(t), Q_d(t), V_w(t))$ in (3.7) would both be insensitive to $V_w(t)$, so that even large fluctuations in the wind speed $V_w(t)$ would be completely absorbed by $\varphi_1(\cdot)$ and $\varphi_2(\cdot)$, producing no fluctuations in the active and reactive powers $P(t)$ and $Q(t)$, which may be unrealistic.

To bypass this limitation, we introduce two second-order linear filters and add two linear terms to (3.6) and (3.7), to produce a modified Hammerstein model depicted in Fig. 3.3c and defined by

$$\begin{bmatrix} \dot{\mu}_{w1}(t) \\ \ddot{\mu}_{w1}(t) \end{bmatrix} = \begin{bmatrix} 0 & 1 \\ -\omega_{n1}^2 & -2\zeta_1\omega_{n1} \end{bmatrix} \begin{bmatrix} \mu_{w1}(t) \\ \dot{\mu}_{w1}(t) \end{bmatrix} + \begin{bmatrix} 0 \\ \omega_{n1}^2 \end{bmatrix} V_w(t), \tag{3.8}$$

$$\begin{bmatrix} \dot{\mu}_{w2}(t) \\ \ddot{\mu}_{w2}(t) \end{bmatrix} = \begin{bmatrix} 0 & 1 \\ -\omega_{n2}^2 & -2\zeta_2\omega_{n2} \end{bmatrix} \begin{bmatrix} \mu_{w2}(t) \\ \dot{\mu}_{w2}(t) \end{bmatrix} + \begin{bmatrix} 0 \\ \omega_{n2}^2 \end{bmatrix} V_w(t), \tag{3.9}$$

$$\dot{P}(t) = -\frac{1}{\tau_1} P(t) + \frac{1}{\tau_1} \varphi_1(P_d(t), Q_d(t), V_w(t)) + \gamma_1(V_w(t) - \mu_{w1}(t)), \tag{3.10}$$

$$\dot{Q}(t) = -\frac{1}{\tau_2} Q(t) + \frac{1}{\tau_2} \varphi_2(P_d(t), Q_d(t), V_w(t)) + \gamma_2(V_w(t) - \mu_{w2}(t)), \tag{3.11}$$

where $\mu_{w1}(t)$ and $\mu_{w2}(t)$ are the filter outputs, $\zeta_1 > 0$ and $\zeta_2 > 0$ are the damping ratios, $\omega_{n1} > 0$ and $\omega_{n2} > 0$ are the natural frequencies, and $\gamma_1 \geq 0$ and $\gamma_2 \geq 0$ are scalar gains. To see the rationale behind (3.8)–(3.11), notice that the second-order linear filters in (3.8) and (3.9) are low-pass filters with unity DC gains. Hence, $\mu_{w1}(t)$ and $\mu_{w2}(t)$ may be seen as short-term averages of $V_w(t)$, which catch up to $V_w(t)$ if it ever approaches constant, and $V_w(t) - \mu_{w1}(t)$ and $V_w(t) - \mu_{w2}(t)$ may be viewed as deviations of $V_w(t)$ from its short-term averages, which fluctuate around zero. It follows that the linear terms $\gamma_1(V_w(t) - \mu_{w1}(t))$ and $\gamma_2(V_w(t) - \mu_{w2}(t))$ in (3.10) and (3.11) enable fluctuations in $V_w(t)$ to induce fluctuations in $P(t)$ and $Q(t)$, bypassing the aforementioned limitation and yielding a feature not possessed by the refined model (3.6) and (3.7). Moreover, because the steady-state values of these terms are zero when $V_w(t)$ is constant, (3.10) and (3.11) also preserve the role of φ_1 and φ_2 as constant-inputs-to-steady-state-outputs maps (see Step 1). Finally, due to the "tuning knobs" ζ_1, ω_{n1}, γ_1, ζ_2, ω_{n2}, and γ_2, (3.8)–(3.11) possess considerable (but not excessive) freedom to mimic the way fluctuations in $V_w(t)$ affect $P(t)$ and $Q(t)$ of the analytical model (3.1) and (3.2). All of these explain the rationale behind (3.8)–(3.11), which we will refer to from now on as the *approximate model*.

The parameters ζ_1, ω_{n1}, γ_1, ζ_2, ω_{n2}, and γ_2 can be identified using the general approach adopted in Step 2, i.e., the so-called prediction-error methods [71]. Indeed, a procedure analogous to the one in Step 2 may be constructed as follows:

Procedure for Step 3

(1) Pick a large $T > 0$.
(2) Pick some specific $(P_d(t), Q_d(t), V_w(t))$ $\forall t \in [0, T]$.
(3) Pick any $x(0)$.
(4) Simulate the analytical model (3.1) and (3.2) from $t = 0$ to $t = T$.
(5) Record $(P(t), Q(t))$ $\forall t \in [0, T]$ as $(P_{an}(t), Q_{an}(t))$.
(6) Pick a large $N_4 \in \mathbb{N}$.
(7) Pick $(\zeta_1^{(i)}, \omega_{n1}^{(i)}, \gamma_1^{(i)}, \zeta_2^{(i)}, \omega_{n2}^{(i)}, \gamma_2^{(i)})$ for $i = 1, 2, \ldots, N_4$.
(8) Use the φ_1 and φ_2 identified in Step 1.
(9) Use the τ_1 and τ_2 identified in Step 2.
(10) Pick any $(\mu_{w1}(0), \dot{\mu}_{w1}(0), P(0), \mu_{w2}(0), \dot{\mu}_{w2}(0), Q(0))$.
(11) Loop over $i = 1, 2, \ldots, N_4$.
(12) Let $(\zeta_1, \omega_{n1}, \gamma_1, \zeta_2, \omega_{n2}, \gamma_2) = (\zeta_1^{(i)}, \omega_{n1}^{(i)}, \gamma_1^{(i)}, \zeta_2^{(i)}, \omega_{n2}^{(i)}, \gamma_2^{(i)})$.
(13) Simulate the model (3.8)–(3.11) from $t = 0$ to $t = T$.
(14) Record $(P(t), Q(t))$ $\forall t \in [0, T]$ as $(P_{ap}^{(i)}(t), Q_{ap}^{(i)}(t))$.
(15) Calculate

$$J_1\left(\zeta_1^{(i)}, \omega_{n1}^{(i)}, \gamma_1^{(i)}\right) = \left(\int_{T_1}^{T} |P_{an}(t) - P_{ap}^{(i)}(t)|^{p_1} dt\right)^{\frac{1}{p_1}}$$

and

$$J_2\left(\zeta_2^{(i)}, \omega_{n2}^{(i)}, \gamma_2^{(i)}\right) = \left(\int_{T_2}^{T} |Q_{an}(t) - Q_{ap}^{(i)}(t)|^{p_2} dt\right)^{\frac{1}{p_2}},$$

where $T_1 \in (0, T)$, $T_2 \in (0, T)$, $p_1 \geq 1$, and $p_2 \geq 1$.
(16) End loop.
(17) Let $(\zeta_1, \omega_{n1}, \gamma_1)$ be the $(\zeta_1^{(i)}, \omega_{n1}^{(i)}, \gamma_1^{(i)})$ that minimizes $J_1(\zeta_1^{(i)}, \omega_{n1}^{(i)}, \gamma_1^{(i)})$, and $(\zeta_2, \omega_{n2}, \gamma_2)$ be the $(\zeta_2^{(i)}, \omega_{n2}^{(i)}, \gamma_2^{(i)})$ that minimizes $J_2(\zeta_2^{(i)}, \omega_{n2}^{(i)}, \gamma_2^{(i)})$.

Remark 3.4 In the above procedure, T may be different from the T in Steps 1 and 2; $P_d(t)$ and $Q_d(t)$ may be, say, staircases, ramps, or from realistic profiles; $V_w(t)$ may be from real data; and $(\zeta_1^{(i)}, \omega_{n1}^{(i)}, \gamma_1^{(i)}, \zeta_2^{(i)}, \omega_{n2}^{(i)}, \gamma_2^{(i)})$ for $i = 1, 2, \ldots, N_4$ represent the search space for the best $(\zeta_1, \omega_{n1}, \gamma_1, \zeta_2, \omega_{n2}, \gamma_2)$.

Note that the three procedures in Steps 1–3 collectively form a parameter identification scheme, which enables sequential determination of all the parameters of the approximate model (3.8)–(3.11) (i.e., φ_1 and φ_2, then τ_1 and τ_2, then the rest).

3.3.2 Approximating the Empirical Model

As pointed out in Sect. 3.2.2, in practice we may be given an empirical model of a WTCS, defined by input–output measurement data of the form (3.3), and asked to design a WFC. Thus, it is desirable that the approximate model (3.8)–(3.11)—with suitable choices of parameters—can also imitate the empirical model (3.3), producing outputs that closely resemble the last two columns of (3.3), when the inputs are from the first three columns. To come up with such suitable choices, reconsider the parameter identification scheme from Steps 1–3. Observe that this scheme is not immediately applicable here because Steps 1 and 2 require constant and staircase inputs $(P_d(t), Q_d(t), V_w(t))$, but with the empirical model (3.3) the inputs $(P_d(i\Delta), Q_d(i\Delta), V_w(i\Delta))$ are whatever that are given. To circumvent this issue, below we modify the scheme, allowing it to handle any given inputs, and label the steps involved Steps 1'–3', to distinguish them from, and to stress their parallel with, Steps 1–3 above. The modification yields the second parameter identification scheme, intended just for the empirical case.

Step 1': Identifying the functions φ_1 and φ_2

Parallel to Step 1, the goal of this Step 1' is to construct a procedure for identifying the functions φ_1 and φ_2, so that the basic model (3.4) and (3.5) in Fig. 3.3a is able to roughly mimic the empirical model (3.3). To do so, observe that the identification of φ_1 (and, similarly, φ_2) can be treated as a curve-fitting problem with domain containing the inputs $(P_d(i\Delta), Q_d(i\Delta), V_w(i\Delta))$ and range containing the output $P(i\Delta)$. Also observe that if we partition the domain into $U_1, U_2, \ldots, U_n$ and write φ_1 as $\varphi_1(P_d, Q_d, V_w) = \sum_{j=1}^{n} \alpha_j \mathbf{1}_{U_j}(P_d, Q_d, V_w)$ where α_j are the parameters and $\mathbf{1}_{U_j}(P_d, Q_d, V_w)$ are the set indicator basis functions (see Chap. 5.4 of [71]), then the optimal α_j in the least-squares sense can be easily computed: each α_j is simply the average of those $P(i\Delta)$ for which $(P_d(i\Delta), Q_d(i\Delta), V_w(i\Delta)) \in U_j$. These observations suggest the following procedure, in which we partition the domain into three-dimensional grids, for simplicity:

Procedure for Step 1'

---

(1) Let

$$\begin{aligned}
P_{d,\max} &= \max_{0 \le i \le D-1} P_d(i\Delta), \\
P_{d,\min} &= \min_{0 \le i \le D-1} P_d(i\Delta), \\
Q_{d,\max} &= \max_{0 \le i \le D-1} Q_d(i\Delta), \\
Q_{d,\min} &= \min_{0 \le i \le D-1} Q_d(i\Delta), \\
V_{w,\max} &= \max_{0 \le i \le D-1} V_w(i\Delta), \\
V_{w,\min} &= \min_{0 \le i \le D-1} V_w(i\Delta).
\end{aligned}$$

(2) Pick a large $N_5 \in \mathbb{N}$.
(3) Let $\delta P_d = \frac{P_{d,\max} - P_{d,\min}}{N_5}$, $\delta Q_d = \frac{Q_{d,\max} - Q_{d,\min}}{N_5}$, and $\delta V_w = \frac{V_{w,\max} - V_{w,\min}}{N_5}$.
(4) Loop over $(j, k, l) \in \{1, 2, \ldots, N_5\}^3$.
(5) Let $I = \Big\{ i \in \{0, 1, \ldots, D-1\} \mid P_d(i\Delta) \in [P_{d,\min} + (j-1)\delta P_d,\ P_{d,\min} + j\delta P_d),\ Q_d(i\Delta) \in [Q_{d,\min} + (k-1)\delta Q_d,\ Q_{d,\min} + k\delta Q_d),\ V_w(i\Delta) \in [V_{w,\min} + (l-1)\delta V_w,\ V_{w,\min} + l\delta V_w) \Big\}$.
(6) If $I \neq \emptyset$, let

$$\varphi_1\Big(P_{d,\min} + (j - \tfrac{1}{2})\delta P_d,\ Q_{d,\min} + (k - \tfrac{1}{2})\delta Q_d,\ V_{w,\min} + (l - \tfrac{1}{2})\delta V_w\Big) = \frac{\sum_{i \in I} P(i\Delta)}{|I|}$$

and

$$\varphi_2\Big(P_{d,\min} + (j - \tfrac{1}{2})\delta P_d,\ Q_{d,\min} + (k - \tfrac{1}{2})\delta Q_d,\ V_{w,\min} + (l - \tfrac{1}{2})\delta V_w\Big) = \frac{\sum_{i \in I} Q(i\Delta)}{|I|},$$

where $|I|$ denotes the cardinality of the set I.
(7) End loop.
(8) Determine φ_1 and φ_2 via interpolation on the (at most) N_5^3 data points obtained.

Step 2': Identifying the parameters τ_1 and τ_2

Unlike going from Step 1 to Step 1' where the procedure undergoes significant changes, only minor modifications are needed to make the procedures in Steps 2 and 3 applicable to the empirical model (3.3) in this Step 2' and the next Step 3'. In particular, the inputs now come from the first three columns of (3.3), p_1 and p_2 now represent the desired ℓ_p norms, and n_1 and n_2 now play the role of T_1 and T_2 in nullifying the impact of the initial states:

Procedure for Step 2'

(1) Rename $(P(i\Delta), Q(i\Delta))$ from the empirical model (3.3) as $(P_{em}(i\Delta), Q_{em}(i\Delta))$.
(2) Pick a large $N_6 \in \mathbb{N}$.
(3) Pick $\tau_1^{(j)} > 0$ and $\tau_2^{(j)} > 0$ for $j = 1, 2, \ldots, N_6$.
(4) Use the φ_1 and φ_2 identified in Step 1'.
(5) Pick any $(P(0), Q(0))$.
(6) Loop over $j = 1, 2, \ldots, N_6$.
(7) Let $\tau_1 = \tau_1^{(j)}$ and $\tau_2 = \tau_2^{(j)}$.
(8) Simulate the refined model (3.6) and (3.7) from $t = 0$ to $t = (D-1)\Delta$.

(9) Record $(P(t), Q(t))\ \forall t \in [0, (D-1)\Delta]$ as $(P_{ap}^{(j)}(t), Q_{ap}^{(j)}(t))$.
(10) Calculate

$$J_1(\tau_1^{(j)}) = \left(\sum_{n=n_1}^{D-1} |P_{em}(n\Delta) - P_{ap}^{(j)}(n\Delta)|^{p_1} \right)^{\frac{1}{p_1}}$$

and

$$J_2(\tau_2^{(j)}) = \left(\sum_{n=n_2}^{D-1} |Q_{em}(n\Delta) - Q_{ap}^{(j)}(n\Delta)|^{p_2} \right)^{\frac{1}{p_2}},$$

where $0 < n_1 < D-1$, $0 < n_2 < D-1$, $p_1 \geq 1$, and $p_2 \geq 1$.
(11) End loop.
(12) Let $\tau_1 = \underset{\tau \in \{\tau_1^{(1)},\ldots,\tau_1^{(N_6)}\}}{\arg\min} J_1(\tau)$ and $\tau_2 = \underset{\tau \in \{\tau_2^{(1)},\ldots,\tau_2^{(N_6)}\}}{\arg\min} J_2(\tau)$.

Step 3': Identifying the parameters $\zeta_1, \omega_{n1}, \gamma_1, \zeta_2, \omega_{n2}$, and γ_2

Procedure for Step 3'

(1) Rename $(P(i\Delta), Q(i\Delta))$ from the empirical model (3.3) as $(P_{em}(i\Delta), Q_{em}(i\Delta))$.
(2) Pick a large $N_7 \in \mathbb{N}$.
(3) Pick $(\zeta_1^{(j)}, \omega_{n1}^{(j)}, \gamma_1^{(j)}, \zeta_2^{(j)}, \omega_{n2}^{(j)}, \gamma_2^{(j)})$ for $j = 1, 2, \ldots, N_7$.
(4) Use the φ_1 and φ_2 identified in Step 1'.
(5) Use the τ_1 and τ_2 identified in Step 2'.
(6) Pick any $(\mu_{w1}(0), \dot{\mu}_{w1}(0), P(0), \mu_{w2}(0), \dot{\mu}_{w2}(0), Q(0))$.
(7) Loop over $j = 1, 2, \ldots, N_7$.
(8) Let $(\zeta_1, \omega_{n1}, \gamma_1, \zeta_2, \omega_{n2}, \gamma_2) = (\zeta_1^{(j)}, \omega_{n1}^{(j)}, \gamma_1^{(j)}, \zeta_2^{(j)}, \omega_{n2}^{(j)}, \gamma_2^{(j)})$.
(9) Simulate the approximate model (3.8)–(3.11) from $t = 0$ to $t = (D-1)\Delta$.
(10) Record $(P(t), Q(t))\ \forall t \in [0, (D-1)\Delta]$ as $(P_{ap}^{(j)}(t), Q_{ap}^{(j)}(t))$.
(11) Calculate

$$J_1(\zeta_1^{(j)}, \omega_{n1}^{(j)}, \gamma_1^{(j)}) = \left(\sum_{n=n_1}^{D-1} |P_{em}(n\Delta) - P_{ap}^{(j)}(n\Delta)|^{p_1} \right)^{\frac{1}{p_1}}$$

and

$$J_2(\zeta_2^{(j)}, \omega_{n2}^{(j)}, \gamma_2^{(j)}) = \left(\sum_{n=n_2}^{D-1} |Q_{em}(n\Delta) - Q_{ap}^{(j)}(n\Delta)|^{p_2} \right)^{\frac{1}{p_2}},$$

where $0 < n_1 < D-1$, $0 < n_2 < D-1$, $p_1 \geq 1$, and $p_2 \geq 1$.
(12) End loop.
(13) Let $(\zeta_1, \omega_{n1}, \gamma_1)$ be the $(\zeta_1^{(j)}, \omega_{n1}^{(j)}, \gamma_1^{(j)})$ that minimizes $J_1(\zeta_1^{(j)}, \omega_{n1}^{(j)}, \gamma_1^{(j)})$, and $(\zeta_2, \omega_{n2}, \gamma_2)$ be the $(\zeta_2^{(j)}, \omega_{n2}^{(j)}, \gamma_2^{(j)})$ that minimizes $J_2(\zeta_2^{(j)}, \omega_{n2}^{(j)}, \gamma_2^{(j)})$.

Remark 3.5 The approximate model (3.8)–(3.11) developed in this section may be viewed as the WTCS$^\theta$ discussed earlier in Sect. 3.1, with $\theta = (\varphi_1, \tau_1, \zeta_1, \omega_{n1}, \gamma_1, \varphi_2, \tau_2, \zeta_2, \omega_{n2}, \gamma_2)$. Also, it may be regarded as satisfying Condition C2 in Sect. 3.1 because it has isolated static nonlinearities and is relatively simple compared to full-blown WTCS models, such as those in Sect. 3.4.

3.4 Model Validation

In this section, we validate via simulation the approximate model developed in Sect. 3.3, showing that it is capable of closely imitating several analytical and empirical WTCS models from the literature and from real data. To enable the validation, we first describe a wind turbine model for General Electric (GE) 3.6 and 1.5 MW turbines, followed by the analytical and empirical WTCS models considered. We then describe the validation settings and results.

3.4.1 *Wind Turbine Model for General Electric* 3.6 *and* 1.5 MW *Turbines*

Reconsider the variable-speed wind turbine with a DFIG, modeled in Sect. 2.2. In particular, recall that the wind turbine's electrical dynamics are modeled by (2.13)–(2.16), its mechanical dynamics are modeled by (2.25), its flux–current relationships are described by (2.5)–(2.8), and its active and reactive power outputs are described by (2.19)–(2.24). Like most mathematical models, this wind turbine model is characterized by a set of parameters, such as the stator and rotor resistances and inductances, rotor moment of inertia, friction coefficient, air density, rotor swept area, and C_p-surface. In what follows, we let this set of parameters take on *two* distinct sets of values.

The first set of values is adopted from [76, 89] and represents a GE 3.6 MW turbine. The values are: $R = 52\,\text{m}$, $\omega_s = 1\,\text{pu}$, $R_s = 0.0079\,\text{pu}$, $R_r = 0.025\,\text{pu}$, $L_s = 4.47937\,\text{pu}$, $L_r = 4.8\,\text{pu}$, $L_m = 4.4\,\text{pu}$, $J = 10.38\,\text{pu}$, $C_f = 0\,\text{pu}$, $V_{w\_\text{base}} = 12\,\text{m/s}$, and the C_p-surface is given by $C_p(\lambda, \beta) = \sum_{i=0}^{4} \sum_{j=0}^{4} \alpha_{ij} \beta^i \lambda^j$, where $\alpha_{44} = 4.9686 \times 10^{-10}$, $\alpha_{43} = -7.1535 \times 10^{-8}$, $\alpha_{42} = 1.6167 \times 10^{-6}$, $\alpha_{41} = -9.4839 \times 10^{-6}$, $\alpha_{40} = 1.4787 \times 10^{-5}$, $\alpha_{34} = -8.9194 \times 10^{-8}$, $\alpha_{33} = 5.9924 \times$

$10^{-6}, \alpha_{32} = -1.0479 \times 10^{-4}, \alpha_{31} = 5.7051 \times 10^{-4}, \alpha_{30} = -8.6018 \times 10^{-4}, \alpha_{24} = 2.7937 \times 10^{-6}$, $\alpha_{23} = -1.4855 \times 10^{-4}$, $\alpha_{22} = 2.1495 \times 10^{-3}$, $\alpha_{21} = -1.0996 \times 10^{-2}$, $\alpha_{20} = 1.5727 \times 10^{-2}$, $\alpha_{14} = -2.3895 \times 10^{-5}$, $\alpha_{13} = 1.0683 \times 10^{-3}$, $\alpha_{12} = -1.3934 \times 10^{-2}$, $\alpha_{11} = 6.0405 \times 10^{-2}$, $\alpha_{10} = -6.7606 \times 10^{-2}$, $\alpha_{04} = 1.1524 \times 10^{-5}$, $\alpha_{03} = -1.3365 \times 10^{-4}$, $\alpha_{02} = -1.2406 \times 10^{-2}$, $\alpha_{01} = 2.1808 \times 10^{-1}$, and $\alpha_{00} = -4.1909 \times 10^{-1}$.

The second set of values is adopted from MATLAB/Simulink and represents a GE 1.5 MW turbine. The values are: $R = 38.5\,\text{m}$, $\omega_s = 1\,\text{pu}$, $R_s = 0.00706\,\text{pu}$, $R_r = 0.005\,\text{pu}$, $L_s = 3.071\,\text{pu}$, $L_r = 3.056\,\text{pu}$, $L_m = 2.9\,\text{pu}$, $J = 10.08\,\text{pu}$, $C_f = 0.01\,\text{pu}$, $V_{w\_\text{base}} = 12\,\text{m/s}$, and the C_p-surface is given by $C_p(\lambda, \beta) = c_1(c_2/\lambda_i - c_3\beta - c_4)e^{-c_5/\lambda_i} + c_6\lambda$, where $1/\lambda_i = 1/(\lambda + 0.08\beta) - 0.035/(\beta^3 + 1)$, and the coefficients are $c_1 = 0.5176, c_2 = 116, c_3 = 0.4, c_4 = 5, c_5 = 21$, and $c_6 = 0.0068$.

3.4.2 WTCS Models from Literature and from Real Data

Next, consider four analytical WTCS models from the literature and one empirical WTCS model from real data, labeled as *WTCS1–WTCS5* and defined as follows:

- **WTCS1** is made up of the GE 3.6 MW turbine model from Sect. 3.4.1 and the controller in Rodriguez-Amenedo et al. [93], which regulates $P(t)$ and $Q(t)$ by

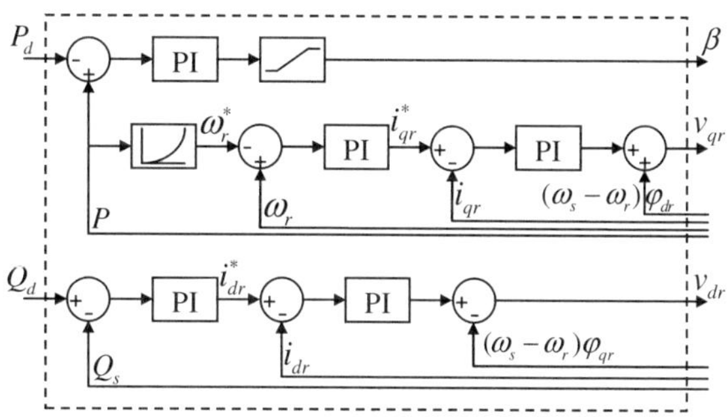

Fig. 3.4 Block diagram of the controller in Rodriguez-Amenedo et al. [93] that yields WTCS1

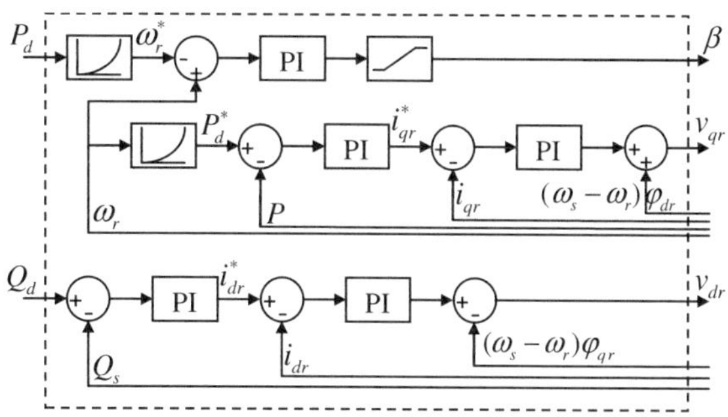

Fig. 3.5 Block diagram of the controller in Fernandez et al. [40] that yields WTCS2

adjusting $\beta(t)$, $v_{qr}(t)$, and $v_{dr}(t)$ using five PI blocks and a power-speed lookup table, as depicted in Fig. 3.4. Note that this controller assumes that the d-axis of the synchronously rotating reference frame is aligned with the stator flux vector, i.e., $(\varphi_{ds}(t), \varphi_{qs}(t)) = (1, 0)$, and that the reactive power is solely coming from the stator, i.e., $Q(t) = Q_s(t)$. For more information about this controller, see [93] and related work [40, 87] (in particular, Figs. 3 of [93], 10 of [87], and 6 and 7 of [40]).

- **WTCS2** is made up of the same GE 3.6 MW turbine model and the controller studied in Fernandez et al. [40] and displayed in Fig. 3.5. Observe that this controller is similar to the one in Rodriguez-Amenedo et al. [93] except that it uses $\omega_r(t)$ to determine $\beta(t)$ in the outer loop and $P(t)$ to determine $v_{qr}(t)$ in the inner loop, whereas the one in [93] does the opposite. For more details about this controller, see [40] and [50] (especially, Figs. 6 and 8 of [40] and 4 of [50]).
- **WTCS3**, unlike WTCS1 and WTCS2, is made up of the smaller GE 1.5 MW turbine model from Sect. 3.4.1 and the nonlinear dual-mode controller in Tang et al. [106], which uses the feedback linearization technique to cancel nonlinearities in the DFIG dynamics, and the gradient descent method to maximize or regulate $P(t)$ and $Q(t)$ including the power factor, as outlined in Fig. 3.6. Notice that this controller assumes instead that the d-axis is aligned with the stator voltage vector, i.e., $(v_{ds}(t), v_{qs}(t)) = (1, 0)$, and that it does not assume $Q(t) = Q_s(t)$.
- **WTCS4** is formed by the mechanical dynamics of the GE 1.5 MW turbine model and the controller in Johnson et al. [59], which is implemented on the Controls Advanced Research Turbine (CART) at the National Renewable Energy Laboratory's (NREL's) National Wind Technology Center (NWTC) and also discussed in [86]. Sketched in Fig. 3.7, this controller maximizes the power capture $T_m(t)\omega_r(t)$ in Region 2 by varying $T_e(t)$ and keeping $\beta(t)$ at its optimum, and prevents the power capture from exceeding the rated value in Region 3 by varying $\beta(t)$ accordingly. In contrast to WTCS1–WTCS3, this WTCS assumes no electrical dynamics and, thus, does not involve $Q_d(t)$ and $Q(t)$, nor $P_d(t)$.
- Finally, **WTCS5**—the only empirical model considered in the model validation—is a black box defined by a set of input–output measurement data taken from an

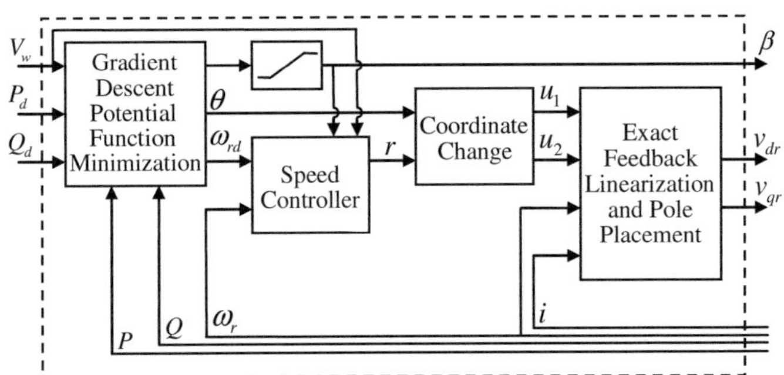

Fig. 3.6 Block diagram of the controller in Tang et al. [106] that yields WTCS3

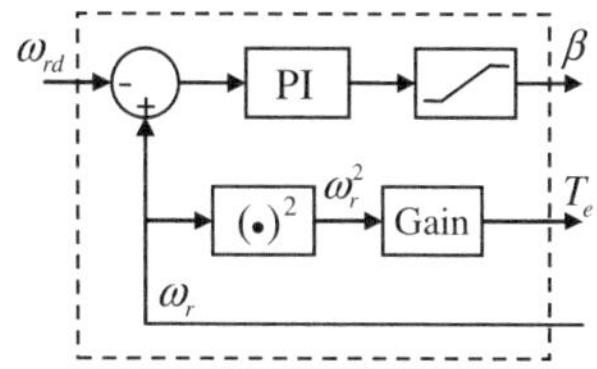

Fig. 3.7 Block diagram of the controller in Johnson et al. [59] that yields WTCS4

actual GE 1.5 MW turbine within a wind farm located in northwest Oklahoma. This set of data has $D = 34{,}208$ data points and was collected over 238 days at a sampling period of $\Delta = 10\,\text{min}$. Moreover, the set of data fits the mold of (3.3), containing the mandatory $V_w(i\Delta)$ and $P(i\Delta)$ for $i = 0, 1, \ldots, D-1$, but not the optional $P_d(i\Delta)$, $Q_d(i\Delta)$, and $Q(i\Delta)$. In order to use this data set for second-level simulation in the sequel, we redefine Δ as $\Delta = 10/24\,\text{min}$, assuming that one-day worth of data were taken over an hour.

3.4.3 Validation Settings

Given WTCS1–WTCS5, suppose now we want to construct, for each WTCSi, an approximate model (3.8)–(3.11) that resembles its behavior. To this end, for each WTCSi, we execute the first parameter identification scheme in Steps 1–3 (if WTCSi is analytical), or the second one in Steps 1'–3' (if it is empirical), to obtain a specific approximate model with specific values of φ_1, τ_1, ζ_1, ω_{n1}, γ_1 as well as φ_2, τ_2, ζ_2, ω_{n2}, γ_2 (if the optional $Q(t)$ is indeed an output of WTCSi).

To evaluate how well the five approximate models imitate WTCS1–WTCS5, we consider 30 different scenarios. For each scenario, we generate inputs $(P_d(t), Q_d(t), V_w(t))$ from $t = 0$ to $t = 3600\,\text{s}$, choosing $P_d(t)$ to be a staircase signal with three random staircase values each lasting 1200 s, $Q_d(t)$ to be such that the desired power factor $P_d(t)/\sqrt{P_d^2(t) + Q_d^2(t)}$ is kept constant at 0.995, and $V_w(t)$ to be an actual wind profile from the aforementioned wind farm. For each WTCSi and each scenario, we simulate both WTCSi and its corresponding approximate model for 3600 s using the *same* inputs $(P_d(t), Q_d(t), V_w(t))$ associated with the scenario, record the outputs $(P(t), Q(t))$ of the two models, and calculate the root-mean-square error (RMSE) in $P(t)$ between the two models after some initial transient. (Obviously, the smaller the RMSE, the better the approximation.)

3.4.4 Validation Results

Figs. 3.8, 3.9, 3.10, 3.11 and 3.12 depict, respectively, the five approximate models and how well they resemble WTCS1–WTCS5. Although the figures have different

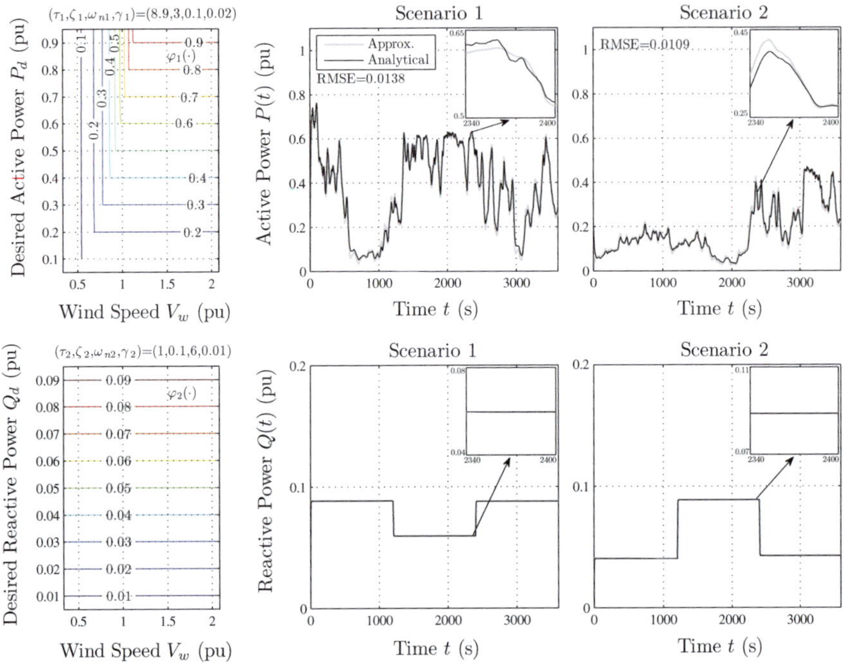

Fig. 3.8 Imitating WTCS1 defined by the GE 3.6 MW turbine and the controller in Rodriguez-Amenedo et al. [93]

sizes and styles, they share the same format: the first row of subplots is associated with the active power; the second row, if present, is associated with the reactive power; the first column displays the identified values of the approximate model parameters $\varphi_1, \tau_1, \zeta_1, \omega_{n1}, \gamma_1, \varphi_2, \tau_2, \zeta_2, \omega_{n2}, \gamma_2$, showing φ_1 and φ_2 as contour plots in Figs. 3.8, 3.9 and 3.10 and as graphs in Figs. 3.11 and 3.12; and the second and third columns each shows, for a selected scenario, the outputs $(P(t), Q(t))$ of WTCSi and those of its corresponding approximate model over 3600 s and over 60 s, the latter in zoom-in windows. Notice that although, in general, φ_1 and φ_2 are functions of (P_d, Q_d, V_w), for WTCS1–WTCS3 they are functions of only (P_d, V_w) or equivalently (Q_d, V_w) (since $Q_d = 0.1 P_d$ in order to obtain a desired power factor of 0.995), and for WTCS4 and WTCS5 they are functions of only V_w (since P_d is not required). This explains why φ_1 and φ_2 can be shown as contour plots and graphs. Also note that for each WTCSi, we show only the outputs for two selected scenarios (as opposed to showing both the inputs and outputs for all the 30 scenarios). Finally, each gray dot in subplot 1 of Fig. 3.12 represents an empirical data point (V_w, P) for WTCS5 and is included just to provide additional insight.

Complementing Figs. 3.8, 3.9, 3.10, 3.11 and 3.12 is Table 3.1, which shows the minimum, maximum, and average RMSE in $P(t)$ between WTCS1–WTCS5 and their corresponding approximate models, taken over all the 30 scenarios. Note that

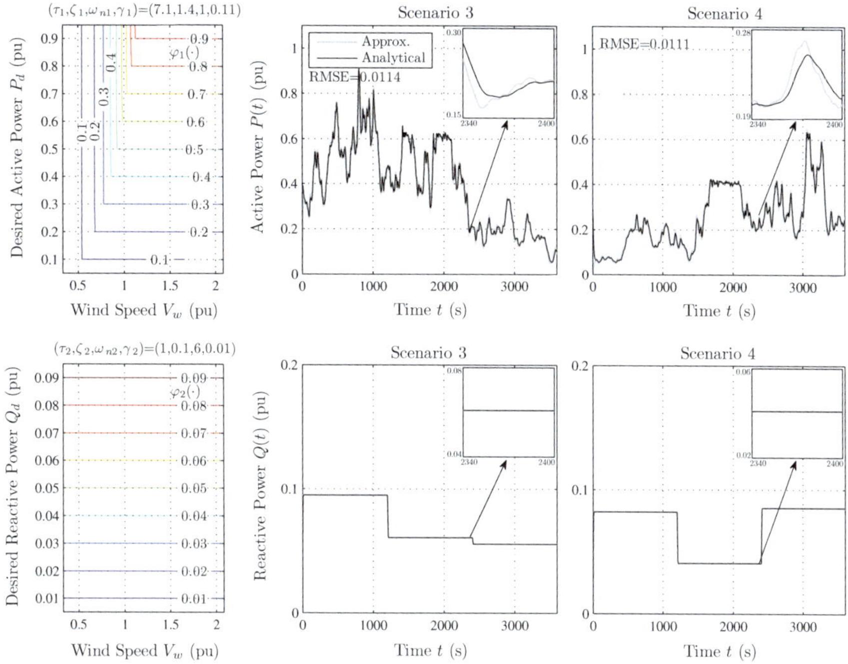

Fig. 3.9 Imitating WTCS2 defined by the GE 3.6 MW turbine and the controller in Fernandez et al. [40]

Table 3.1 carries significantly more weight than Figs. 3.8, 3.9, 3.10, 3.11 and 3.12 because it reports the results of 30 scenarios, whereas Figs. 3.8, 3.9, 3.10, 3.11 and 3.12 each reports only the results of two selected scenarios, albeit with more information on each scenario. To get a sense of what the numbers in Table 3.1 mean, one may refer to Figs. 3.8, 3.9, 3.10, 3.11 and 3.12, which also state the RMSEs of the $P(t)$ curves for the few selected scenarios.

Observe from Figs. 3.8, 3.9, 3.10, 3.11 and 3.12 and Table 3.1 that while the proposed approximate model is not without error, the magnitude of which is generally

Table 3.1 Minimum, maximum, and average root-mean-square error (RMSE) in $P(t)$ taken over 30 scenarios for each WTCSi

	Minimum	Maximum	Average
WTCS1	0.0103	0.0178	0.0134
WTCS2	0.0111	0.0164	0.0137
WTCS3	0.0111	0.0237	0.0163
WTCS4	0.0056	0.0088	0.0072
WTCS5	0.0223	0.0578	0.0320

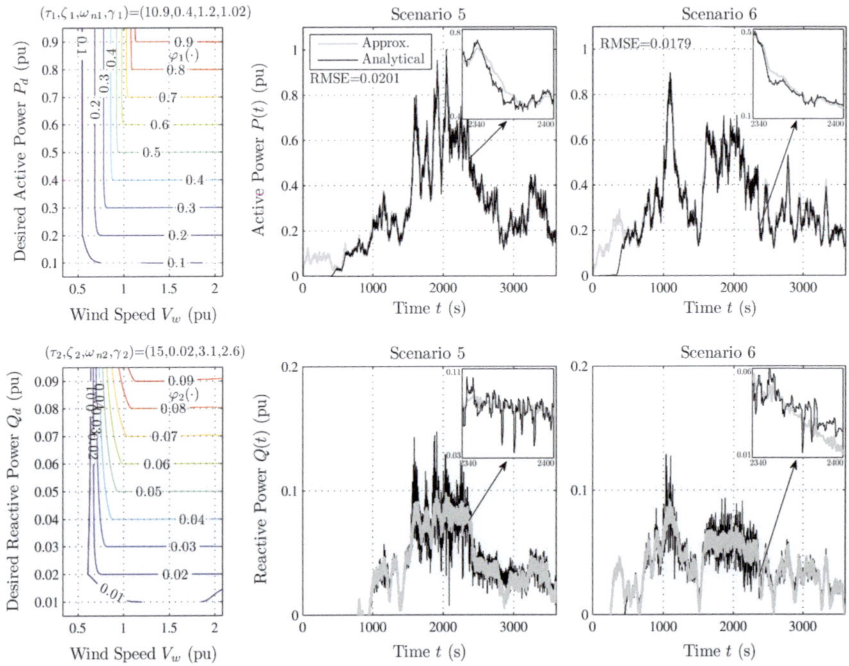

Fig. 3.10 Imitating WTCS3 defined by the GE 1.5 MW turbine and the controller in Tang et al. [106]

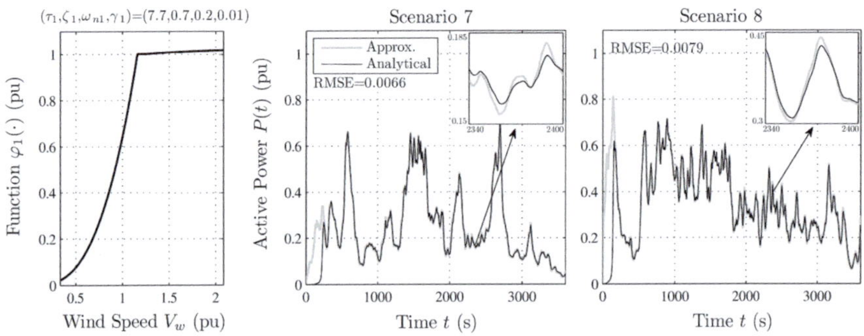

Fig. 3.11 Imitating WTCS4 defined by the mechanical dynamics of the GE 1.5 MW turbine and the controller in Johnson et al. [59]

very small, sometimes even negligible, across all WTCSs and all scenarios. In particular, it is able to produce the right "peaks" and "valleys" at the right moments in all the 3600-s subplots and the 60-s zoom-in windows—except for the first 500 s in Fig. 3.10 and first 200 s in Fig. 3.11, which may be attributed to the approximate and analytical models having different initial states and, hence, different initial transients (i.e., the warm-up periods for the models). These encouraging observations

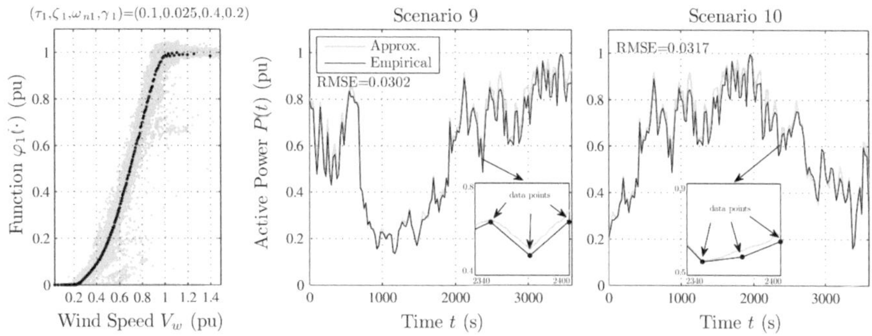

Fig. 3.12 Imitating WTCS5 defined by real data from an Oklahoma wind farm

validate the approximate model in Fig. 3.3c, demonstrating its ability to closely replicate the behaviors of the five fairly different analytical and empirical WTCS models considered.

Remark 3.6 The validation in this section may be thought of as verifying Conditions C1 and C3 in Sect. 3.1.

3.5 Concluding Remarks

In this chapter, we have presented a simple approximate model, which tries to mimic the active and reactive power dynamics of generic analytical and empirical WTCS models, along with two parameter identification schemes, which determine the approximate model parameters in both cases. We have also demonstrated through simulation the ability of the approximate model in resembling several different analytical and empirical WTCS models from the literature and from real data. The results suggest that the approximate model is a compelling candidate, based on which one may design and analyze a second-to-minute-timescale supervisory wind farm controller using a variety of control techniques.

Chapter 4
Model Predictive and Adaptive Control of Wind Farm Active Power Output

Abstract In this chapter, we introduce a novel supervisory wind farm controller that enables the active power output of a wind farm to accurately and smoothly track a desired reference provided by a power grid operator. Developed based on the approximate wind turbine control systems model from Chap. 3, this wind farm controller has a two-loop architecture, consisting of an outer feedback loop and an inner one. The outer feedback loop contains a model predictive controller, which uses various forecasts and feedbacks to iteratively compute a set of desired power trajectories, so that the deterministic tracking accuracy of the wind farm power output on a receding horizon is optimized. In contrast, the inner feedback loop contains an adaptive controller, which uses estimated wind speed characteristics to adaptively tune a set of proportional controller gains, so that the stochastic smoothness of the wind farm power output on a shorter timescale is optimized. We also present a series of simulation studies that illustrate the salient features of the wind farm controller, including its ability to exploit forecast availability, design freedom, and wind speed correlation.

4.1 Introduction

Wind farms that can produce power outputs that accurately and smoothly track desired references from a power grid operator despite the variability and intermittency of wind, are important. This ability allows them to be treated more or less as "controllable" generation resources, similar to conventional power plants. It also reduces their reliance on expensive ancillary services, leading to more economic operation. Moreover, with the increasing penetration of wind power in the energy portfolios of many countries, this ability becomes especially critical to power system reliability. Therefore, a *wind farm controller* (WFC) that can provide a wind farm with such accurate and smooth tracking ability, is valuable.

To date, a number of WFCs have been proposed in the literature [32, 40, 50, 70, 72, 74, 81, 93, 96, 101, 103, 104, 107] based on different models and techniques and for different purposes. For example, [50, 93] each proposes a hierarchical, supervisory WFC that determines the active and reactive power setpoints of every turbine so that

J.N. Jiang et al., *Control and Operation of Grid-Connected Wind Farms*,
Advances in Industrial Control, DOI 10.1007/978-3-319-39135-9_4

the wind farm power output is regulated around a desired reference; [32, 81] each considers a static turbine model and introduces an optimization-based approach for designing WFCs that can respond to grid operator requests in an optimal fashion; [96] utilizes fuzzy neural networks to design a WFC that adjusts the blade pitch angle of every turbine so that variations in the wind farm power output are reduced; and [72] suggests a supervisory WFC that uses either an external energy storage device, or a power reserve achieved through part-loading of some turbines, to make the wind farm power output smooth. As additional examples, [107] devises a proportional-integral regulator-based WFC that aims at managing the wind farm reactive power for secondary voltage control; [40] carries out simulation studies that compare the performance of several existing WFCs equipped with power regulation capabilities; [70] presents a WFC that controls the blade pitch angles in unison so that the wind farm power output can follow a desired reference while the damping power can be supplied as an ancillary service; and [74] develops a distributed learning WFC that maximizes the wind farm power output without having to explicitly model the aerodynamic interactions among the turbines.

Although the existing WFCs possess a number of desirable attributes, they do not seem to take advantage of the following three typical characteristics of a wind farm:

B1. *Forecast availability*: In addition to providing a wind farm with a desired reference that represents the amount of power the wind farm should generate, a power grid operator often has access to a several-hours-ahead forecast of the desired reference. Moreover, while wind speeds are generally difficult to predict, reliable forecasts of their average values up to a few hours often can be made [18, 45, 64, 78]. Conceivably, incorporating such forecasts into a WFC may significantly enhance its performance.

B2. *Design freedom*: Wind farm power control in the present context is, mathematically, a problem with a large design freedom: as far as the power grid operator is concerned, what matters is that the *sum* of all the turbine power outputs is equal to the desired reference. Thus, if a wind farm has N turbines, $N - 1$ degrees of freedom would remain upon satisfying this equality constraint. Conceivably, such freedom may be used by a WFC to achieve other secondary goals, such as reducing abrupt changes to the control signals and, hence, to the turbine parts, benefiting their lifetime.

B3. *Wind speed correlation*: Because of their geographical proximity, turbines in a wind farm inevitably experience wind speeds that are correlated, with correlation that generally changes slowly over time due to gradual changes in wind directions, weather conditions, and turbine yaw angles. It is conceivable that exploiting such correlation in a WFC may substantially improve its performance.

This chapter is devoted to the design and analysis of a WFC for active power control that takes advantage of Characteristics B1–B3. More specifically, we consider a *wind farm control system* (WFCS) comprising a wind farm and its WFC, as shown in Fig. 4.1. The wind farm consists of N *wind turbine control systems* (WTCSs), where each WTCS $i \in \{1, 2, \ldots, N\}$ is made up of a wind turbine and its controller. Accompanying the WFCS is a wind speed block that produces wind speeds

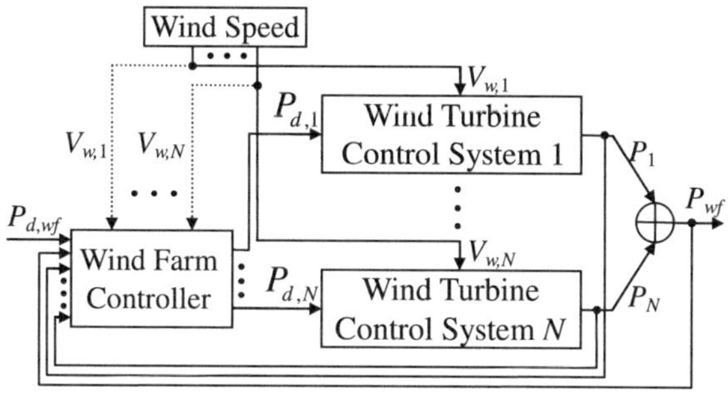

Fig. 4.1 Block diagram of a wind farm control system for active power control

$V_{w,1}, \ldots, V_{w,N}$, where each $V_{w,i}$ affects WTCS i and can be (roughly) measured by the WFC. In addition to accepting $V_{w,i}$, each WTCS i accepts a desired active power reference $P_{d,i}$ from the WFC and produces an active power output P_i. Meanwhile, the WFC uses a desired wind farm active power output $P_{d,wf}$ from the power grid operator, noisy measurements of $V_{w,1}, \ldots, V_{w,N}$, and feedback of $P_1, \ldots, P_N$ to calculate the control inputs $P_{d,1}, \ldots, P_{d,N}$ that are sent to the WTCSs, so that the wind farm active power output $P_{wf} \triangleq \sum_{i=1}^{N} P_i$ closely tracks $P_{d,wf}$.

As will be detailed in Sects. 4.3–4.5, the WFC consists of an outer feedback loop and an inner one. The outer feedback loop comprises a *model predictive controller* that optimizes the deterministic tracking accuracy of P_{wf} on a receding horizon, while the inner one comprises an *adaptive controller* that optimizes the stochastic smoothness of P_{wf} on a shorter timescale.

To achieve its goal, the model predictive controller uses a forecast of $P_{d,wf}$, noisy measurements of $V_{w,1}, \ldots, V_{w,N}$, and feedback of $P_1, \ldots, P_N$ to iteratively compute a set of intermediate power references that drive the adaptive controller. Although model predictive control has been successfully utilized in many applications [21, 94], we believe its use in wind farm power control in this context has not been reported. This approach appears to be natural for the problem at hand as it allows the WFC to make use of the forecast availability and design freedom mentioned in Characteristics B1 and B2.

Likewise, to achieve its goal, the adaptive controller, which is of the *self-tuning regulator* [9] type, uses estimated wind speed characteristics (e.g., correlation) from measurements to adaptively tune the gains of a set of N decoupled proportional controllers that calculate the control inputs $P_{d,1}, \ldots, P_{d,N}$. While adaptive control has found widespread applications [9], we are unaware of its use in wind farm power control. This approach seems to be particularly suitable for the problem in consideration because, together with stochastic linear systems theory, it enables the WFC to exploit the wind speed correlation mentioned in Characteristic B3 in both design and analysis.

Finally, in order to demonstrate the effectiveness of the WFC developed, we carry out several sets of simulations that test its model predictive controller in isolation, its adaptive controller in isolation, and the overall WFC.

4.2 Wind Farm Model and Problem Formulation

Recall from Fig. 4.1 that the WFCS contains N WTCSs and a WFC and is accompanied by a wind speed block. In Sects. 4.2.1 and 4.2.2, we model the wind speed block and the N WTCSs, which, together, form a wind farm model. In Sect. 4.2.3, we formulate the control problem to be addressed.

4.2.1 *Wind Speed Model*

Observe from Fig. 4.1 that the wind speed block produces N wind speeds $V_{w,1}(t), \ldots, V_{w,N}(t)$ entering WTCS 1 through N. For each $i \in \{1, 2, \ldots, N\}$, we assume that $V_{w,i}(t)$ can be decomposed as

$$V_{w,i}(t) = \overline{V}_{w,i}(t) + \widetilde{V}_{w,i}(t), \tag{4.1}$$

where $t \geq 0$ denotes time, $\overline{V}_{w,i}(t) > 0$ represents the slow, average component of $V_{w,i}(t)$ on a minute-to-hour timescale, and $\widetilde{V}_{w,i}(t) \in \mathbb{R}$ represents the fast, deviation-from-average component of $V_{w,i}(t)$ on a millisecond-to-second timescale. The slow components $\overline{V}_{w,1}(t), \ldots, \overline{V}_{w,N}(t)$ are assumed to be deterministic and specified by empirical data (e.g., historical hourly weather data) or test signals (e.g., step, ramp, sinusoidal). In contrast, the fast components $\widetilde{V}_{w,1}(t), \ldots, \widetilde{V}_{w,N}(t)$ are assumed to be stochastic and specified by

$$\begin{bmatrix} \widetilde{V}_{w,1}(t) \\ \vdots \\ \widetilde{V}_{w,N}(t) \end{bmatrix} = \mathcal{L}^{-1}\{G_w(s)\} * w(t), \tag{4.2}$$

where $*$ denotes the convolution operator, $\mathcal{L}^{-1}$ denotes the inverse Laplace transform, $w(t) \in \mathbb{R}^{N_w}$ is a stationary, zero-mean white Gaussian random process with autocovariance function $\mathrm{E}\{w(t)w^T(\tau)\} = W\delta(t-\tau)$, $G_w(s)$ is an N-by-N_w asymptotically stable transfer function matrix, E denotes the expectation operator, $W \in \mathbb{R}^{N_w \times N_w}$ is a symmetric positive definite covariance matrix, and δ denotes the Dirac delta function. Note that due to (4.2), $\widetilde{V}_{w,1}(t), \ldots, \widetilde{V}_{w,N}(t)$ are stationary, zero-mean colored Gaussian random processes. Also, because $\widetilde{V}_{w,i}(t)$ is Gaussian, $V_{w,i}(t)$ may be negative with a small probability despite $\overline{V}_{w,i}(t)$ being positive. For simplicity, however, we will allow that in this chapter. Also note that in reality, $\widetilde{V}_{w,1}(t), \ldots, \widetilde{V}_{w,N}(t)$ may be nonstationary due to changes in wind directions, weather conditions, and turbine yaw angles. Such changes, however, are usually very slow, so that they can be considered stationary. Moreover, although $G_w(s)$ above is a general transfer function matrix, we will assume a specific $G_w(s)$ in Sect. 4.5 for concreteness. Finally, the top portion of Fig. 4.2 represents the wind speed model as described by (4.1) and (4.2).

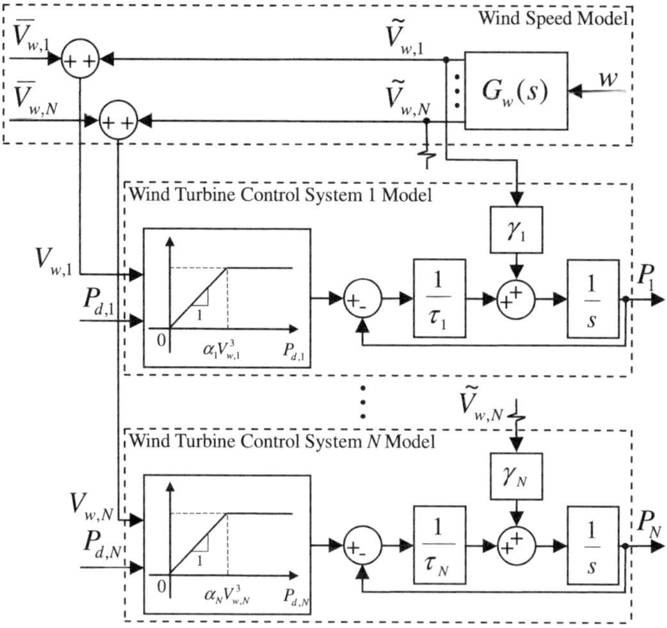

Fig. 4.2 Wind farm model formed by the wind speed model from Sect. 4.2.1 and N copies of the WTCS model from Sect. 4.2.2

4.2.2 Wind Turbine Control System Model

As discussed in Sect. 3.1, a WTCS comprising a wind turbine and its controller is a fairly complex nonlinear dynamical system, which makes the design and analysis of a sophisticated WFC nontrivial. This motivates the need to build a suitably simplified WTCS model, which has been accomplished in Chap. 3. More specifically, we have constructed in Sect. 3.3 a simple approximate model described by (3.8)–(3.11) and illustrated in Fig. 3.3c. We have also validated this approximate model in Sect. 3.4, showing via Figs. 3.8–3.12 and Table 3.1 that it is capable of closely imitating the active and reactive power dynamics of several different WTCS models from the literature [40, 59, 93, 106] and from real data taken from an Oklahoma wind farm. In this chapter, we assume that each WTCS $i \in \{1, 2, \ldots, N\}$ is modeled by the said approximate model. In addition, we consider only the active power and assume that the reactive power is adjusted accordingly by some turbine-level control, so that a constant power factor is maintained. With these assumptions, we will mostly omit the term "active" in the sequel, and write the dynamics of each WTCS $i \in \{1, 2, \ldots, N\}$ using (3.10) as

$$\dot{P}_i(t) = -\frac{1}{\tau_i} P_i(t) + \frac{1}{\tau_i}\varphi_i(P_{d,i}(t), V_{w,i}(t)) + \gamma_i(V_{w,i}(t) - \mu_{w,i}(t)), \tag{4.3}$$

where, again, $P_i(t) \in \mathbb{R}$ is the power output of WTCS i, $P_{d,i}(t) \geq 0$ is its desired power reference, $\mu_{w,i}(t)$ may be seen as a short-term average of $V_{w,i}(t)$, φ_i is a static nonlinear function, $\tau_i > 0$ is the (dominant) time constant, and $\gamma_i \geq 0$ is a scalar gain.

Remark 4.1 Note that there is a notational difference between (3.8)–(3.11) and (4.3): because (3.8)–(3.11) consider only *one* turbine but both *active* and *reactive* powers, there is no need to index the turbine, and the subscripts 1 and 2 in (3.8)–(3.11) are used to represent active and reactive powers, respectively (see Remark 3.2). In contrast, because (4.3) applies to *every* turbine i but considers only *active* power, there is a need to index each turbine i with a subscript i, but no need to distinguish between active and reactive powers.

Although WTCS model (4.3) is quite simple, it is possible to further simplify it in two ways as follows: first, since both $\mu_{w,i}(t)$ defined in (3.8) and appearing in (4.3) and $\overline{V}_{w,i}(t)$ defined in (4.1) play similar roles as short-term averages of $V_{w,i}(t)$, we view them as the same quantity, i.e.,

$$\mu_{w,i}(t) \equiv \overline{V}_{w,i}(t),$$

so that $V_{w,i}(t) - \mu_{w,i}(t)$ in (4.3) may be replaced by $\widetilde{V}_{w,i}(t)$ in (4.1). This replacement means that the wind speed model from Sect. 4.2.1, instead of the $\mu_{w,i}(t)$ dynamics (3.8), is used to define the rightmost term in (4.3). Second, as it turns out, the static nonlinear function $\varphi_i(P_{d,i}, V_{w,i})$ in (4.3) can be accurately approximated by a saturation function of $P_{d,i}$ with an upper limit of $\alpha_i V_{w,i}^3$ and a lower limit of 0, i.e.,

$$\varphi_i(P_{d,i}, V_{w,i}) \approx \text{sat}_0^{\alpha_i V_{w,i}^3}(P_{d,i}), \quad \forall (P_{d,i}, V_{w,i}),$$

where $\text{sat}_a^b(x) \triangleq \max\{\min\{x, b\}, a\}$, and $\alpha_i > 0$ represents a unit conversion factor. The upper limit of $\alpha_i V_{w,i}^3$ corresponds to the physics-based, cubic relationship between wind speed and its power, while the lower limit of 0 is due to power output being nonnegative in normal operating regimes. Figure 4.3 shows that this approximation is indeed highly accurate for at least three analytical WTCS models from the literature (see Sect. 3.4 for more details), and we believe that majority of well-designed WTCSs available today exhibit similar attributes. Consequently, the function $\varphi_i(P_{d,i}(t), V_{w,i}(t))$ in (4.3) may be replaced by the function $\text{sat}_0^{\alpha_i V_{w,i}^3(t)}(P_{d,i}(t))$.

With the above two modifications to (4.3), we obtain a further simplified model given by

$$\dot{P}_i(t) = -\frac{1}{\tau_i} P_i(t) + \frac{1}{\tau_i}\, \text{sat}_0^{\alpha_i V_{w,i}^3(t)}(P_{d,i}(t)) + \gamma_i \widetilde{V}_{w,i}(t), \tag{4.4}$$

which will be used in all subsequent WFC design and analysis. We note that WTCS model (4.4) is not without limitations. Because of its high simplicity, it neglects many details of the turbine and controller dynamics, including the turbine electrical

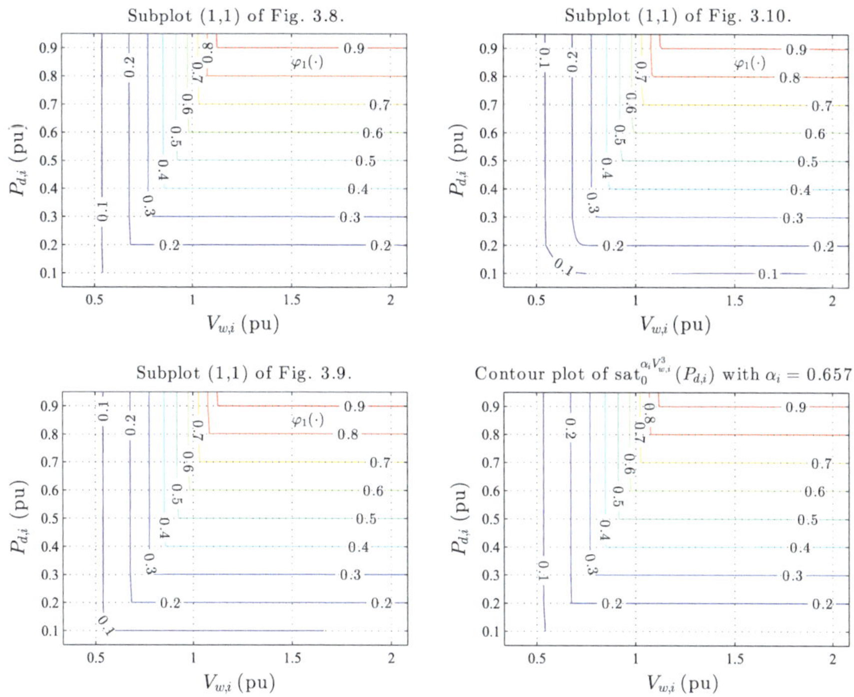

Fig. 4.3 Comparison between the contour plots of the functions $\varphi_1(\cdot)$ from subplot (1, 1) of Figs. 3.8–3.10 and the contour plot of the function $\mathrm{sat}_0^{\alpha_i V_{w,i}^3}(P_{d,i})$ with $\alpha_i = 0.657$ shows that the latter is an excellent approximation of the former

dynamics and flexible modes, and captures only its first-order, dominant transient behavior via the time constant τ_i. Nevertheless, WTCS model (4.4) makes physical sense: for example, when the wind is strong enough, i.e., $\alpha_i V_{w,i}^3(t) > P_{d,i}(t)$, the saturation function does not come into play, so that the power output $P_i(t)$ tends to track the desired power reference $P_{d,i}(t)$, causing the WTCS to operate in the PR mode. Otherwise, i.e., $\alpha_i V_{w,i}^3(t) \le P_{d,i}(t)$, $P_i(t)$ tends to track the theoretical limit $\alpha_i V_{w,i}^3(t)$, causing it to operate in the MPT mode. Also notice that the "disturbance" term $\gamma_i \widetilde{V}_{w,i}(t)$ enables fast fluctuations in $V_{w,i}(t)$ to induce fluctuations in $P_i(t)$, making the WTCS dynamics more realistic. Finally, notice that (4.4) is represented in the bottom portion of Fig. 4.2 and describes the dynamics of each WTCS $i \in \{1, 2, \ldots, N\}$ in Fig. 4.1.

4.2.3 Problem Statement

Having introduced the wind farm model formed by wind speed model (4.1) and (4.2) and N copies of WTCS model (4.4), we next address the following control problem: design a WFC (i.e., the WFC block in Fig. 4.1) that determines the desired power references $P_{d,i}(t)$'s based on feedback of the power outputs $P_i(t)$'s, noisy measurements of the wind speeds $V_{w,i}(t)$'s, a desired wind farm power output $P_{d,wf}(t)$, and a several-hours-ahead forecast of $P_{d,wf}(t)$, so that the wind farm power output $P_{wf}(t)$ accurately and smoothly tracks $P_{d,wf}(t)$. The WFC may use values of the turbine-dependent parameters (i.e., α_i's, τ_i's, and γ_i's) and the value of N. More importantly, the WFC should take advantage of Characteristics B1–B3 stated in Sect. 4.1.

4.3 Wind Farm Controller Architecture

In this and the next two sections, we develop a WFC that solves the aforementioned problem. This section presents the rationale behind the WFC architecture and introduces the WFC block diagram, while the next two provide the technical details.

4.3.1 Rationale Behind Architecture

Recall from Sect. 4.2.3 that the WFC has the following two goals:

G1. Make $P_{wf}(t)$ accurately track $P_{d,wf}(t)$.
G2. Make $P_{wf}(t)$ as smooth as possible despite the wind speed fluctuations.

Notice that an accurate $P_{wf}(t)$ is not necessarily smooth, and vice versa, as illustrated in Fig. 4.4. Thus, Goals G1 and G2 are somewhat independent. Also note from Fig. 4.1 that $P_{wf}(t) = \sum_{i=1}^{N} P_i(t)$. Hence, Goals G1 and G2 are perfectly achieved if for all $t \geq 0$,

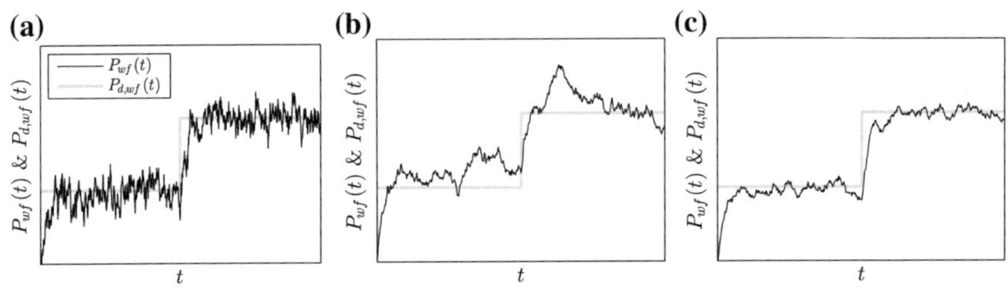

Fig. 4.4 Illustration of the difference between tracking accuracy and smoothness. **a** Accurate but not smooth. **b** Smooth but not accurate. **c** Accurate and smooth

$$\sum_{i=1}^{N} P_i(t) = P_{d,wf}(t). \tag{4.5}$$

Imagine, for a moment, that we may freely specify the values of the N power outputs $P_i(t)$'s in (4.5). Then, as mentioned in Characteristic B2 of Sect. 4.1, (4.5) represents an equality constraint, the satisfaction of which leaves us with $N - 1$ degrees of freedom, which are abundant given that N is usually large in a wind farm. Obviously, there are two opposing ways to handle such freedom: one may simply ignore it and choose, say, $P_i(t) = \frac{1}{N} P_{d,wf}(t)$ for every $i \in \{1, 2, \ldots, N\}$ (i.e., uniformly distribute $P_{d,wf}(t)$ to every turbine), or one may opportunistically select the $P_i(t)$'s to meet some secondary goals, in addition to satisfying (4.5). Clearly, the latter is preferred over the former, provided that meaningful secondary goals can be defined.

To this end, we make three observations: first, note from Fig. 4.1 that it is the inputs $P_{d,i}(t)$'s of the N WTCSs, rather than their outputs $P_i(t)$'s, which may be freely specified. Moreover, dramatic jumps in the $P_{d,i}(t)$'s may lead to rapid changes in the turbine states, possibly exciting some high frequency unmodeled turbine dynamics and/or causing undesirable mechanical vibrations. Second, as mentioned in Characteristic B1, although the $V_{w,i}(t)$'s are generally difficult to predict—especially their fast components $\widetilde{V}_{w,i}(t)$'s—their slow components $\overline{V}_{w,i}(t)$'s do exhibit predictable trends, at least for several minutes and even up to a few hours. Third, as also mentioned in Characteristic B1, a several-hours-ahead forecast of $P_{d,wf}(t)$ is typically available from the power grid operator.

The above three observations suggest that the $P_{d,i}(t)$'s may be determined by solving an optimization problem, in which the cost function is a sum of a tracking accuracy term (motivated by (4.5)) and a control effort term (motivated by the first observation), taken over a finite horizon into the future (motivated by the second and third observations). Since forecasts of $P_{d,wf}(t)$ and the $\overline{V}_{w,i}(t)$'s may be revised for better accuracy as time elapses, repeatedly solving the said optimization problem over a receding or moving time horizon and applying only the initial portion of the optimal solution $P_{d,i}(t)$'s yield a strategy that incorporates not only the three observations toward satisfying (4.5), but also the availability of revised forecasts. Given that this is the philosophy of *model predictive control* [21, 94], we refer to the strategy as a *model predictive controller*. Notice that while a forecast of $P_{d,wf}(t)$ is readily available from the power grid operator, forecasts of the $\overline{V}_{w,i}(t)$'s may not be. Therefore, an additional, forecasting block within the model predictive controller may be needed.

Up to this point, we have provided the justification for having the model predictive controller, which computes the $P_{d,i}(t)$'s that drive the N WTCSs. This controller, however, has two limitations: first, the optimization problem of determining the $P_{d,i}(t)$'s must make use of WTCS model (4.4) from a current time $t_0 \geq 0$ to a future time $t_0 + T$, where $T > 0$ is the optimization horizon. Since the $\widetilde{V}_{w,i}(t)$'s in (4.4) are not known for all $t \in [t_0, t_0 + T]$, and since they are fast-changing, zero-mean, and Gaussian, a simple and reasonable way to use (4.4) in the optimization is to disregard the $\widetilde{V}_{w,i}(t)$'s. Thus, the model predictive controller is not expected to handle the

$\widetilde{V}_{w,i}(t)$'s. Second, if we consider discrete-time model predictive control and divide the optimization horizon T into $K \geq 1$ time slots, the optimization problem would have dimension NK. Since N is usually large, for the problem to be practically feasible, K cannot be very large. Since a sufficiently large T is often desirable to take advantage of the available forecasts, this means that the time slot duration T/K is typically large. As a result, the model predictive controller is expected to make $P_{wf}(t)$ accurately track $P_{d,wf}(t)$ on a longer timescale, but not expected to make $P_{wf}(t)$ smooth on a shorter one.

The above two limitations suggest that although the model predictive controller may be used to achieve Goal G1, it cannot be used to achieve Goal G2 of making $P_{wf}(t)$ smooth on a shorter timescale by rejecting the fast-changing $\widetilde{V}_{w,i}(t)$'s in (4.4). To achieve this Goal G2, the model predictive controller needs help. One way to help is to insert, at the point where the computed $P_{d,i}(t)$'s are supposed to enter the WTCSs (see Figs. 4.1 and 4.2), a feedback controller that uses the computed $P_{d,i}(t)$'s (which we will rename as $P_{d,i}^*(t)$'s to avoid ambiguity) and feedback of the $P_i(t)$'s to calculate the "corrected" $P_{d,i}(t)$'s, which actually enter the WTCSs. Due to its location, this feedback controller and the model predictive controller may be considered as forming an *inner feedback loop* and an *outer feedback loop* of a WFC, respectively. In this chapter, we let this feedback controller be a set of N decoupled proportional controllers for simplicity, and note that other, more sophisticated choices are certainly possible. Moreover, we precede each proportional controller with a feedforward gain, which will be shown in Sect. 4.5.2 to eliminate steady-state error. Furthermore, we precede each feedforward gain with a reference model, which converts $P_{d,i}^*(t)$ into an intermediate power reference denoted as $P_i^*(t)$. The $P_i^*(t)$'s are to be used by the proportional controllers to reject the $\widetilde{V}_{w,i}(t)$'s and make $P_{wf}(t)$ smooth, achieving Goal G2.

With the insertion of the proportional controllers for making $P_{wf}(t)$ smooth, comes the questions of how to define smoothness, and how to choose the proportional controller gains. In this chapter, we define the smoothness of $P_{wf}(t)$ as its steady-state variance, so that by neglecting the saturation function in (4.4) and by leveraging stochastic linear systems theory, the smoothness of $P_{wf}(t)$ may be expressed as an explicit function of the proportional controller gains and the wind speed parameters. It follows that if the wind speed parameters are known, the proportional controller gains may be chosen to optimize smoothness, exploiting Characteristic B3. Unfortunately however, the wind speed parameters are not readily known. Hence, an additional block is needed to provide estimates of such parameters. Furthermore, since the values of such parameters generally change slowly over time due to gradual changes in wind speed characteristics, this additional block should periodically update its estimates of the wind speed parameters, and the proportional controller gains should be periodically optimized as well. Given that this is the idea of a *self-tuning regulator* in *adaptive control* [9], we refer to the overall scheme on the inner feedback loop as an *adaptive controller*.

4.3.2 Controller Block Diagram

Putting together the above line of thoughts, we obtain a WFC containing a model predictive controller on an outer feedback loop and an adaptive controller on an inner one, as depicted in Fig. 4.5. Note that Fig. 4.5 may be regarded as showing the internal details of the WFC block in Fig. 4.1.

As shown on the left of Fig. 4.5, the model predictive controller is made up of three blocks, namely, Forecasts of Slow Wind Speed Components, Optimization of Desired Power Trajectories, and Reference Models, which operate as follows: based on noisy measurements of the wind speeds $V_{w,i}(t)$'s, the Forecasts of Slow Wind Speed Components block produces forecasts $\widehat{V}_{w,i}(t)$'s of their slow components $\overline{V}_{w,i}(t)$'s. The $\widehat{V}_{w,i}(t)$'s, along with feedback of the power outputs $P_i(t)$'s, the desired wind farm power output $P_{d,wf}(t)$, and the forecast of $P_{d,wf}(t)$, are used by the Optimization of Desired Power Trajectories block to produce the optimal desired power trajectories $P^*_{d,i}(t)$'s. The $\widehat{V}_{w,i}(t)$'s and $P^*_{d,i}(t)$'s, in turn, are used by the Reference Models block to produce the intermediate power references $P^*_i(t)$'s that drive the adaptive controller.

Likewise, as shown on the right of Fig. 4.5, the adaptive controller is made up of four blocks, namely, Estimation of Wind Speed Parameters, Optimization of Proportional Controller Gains, Feedforward Gains, and Proportional Controllers, which operate as follows: based on noisy measurements of the wind speeds $V_{w,i}(t)$'s, the Estimation of Wind Speed Parameters block produces an estimate

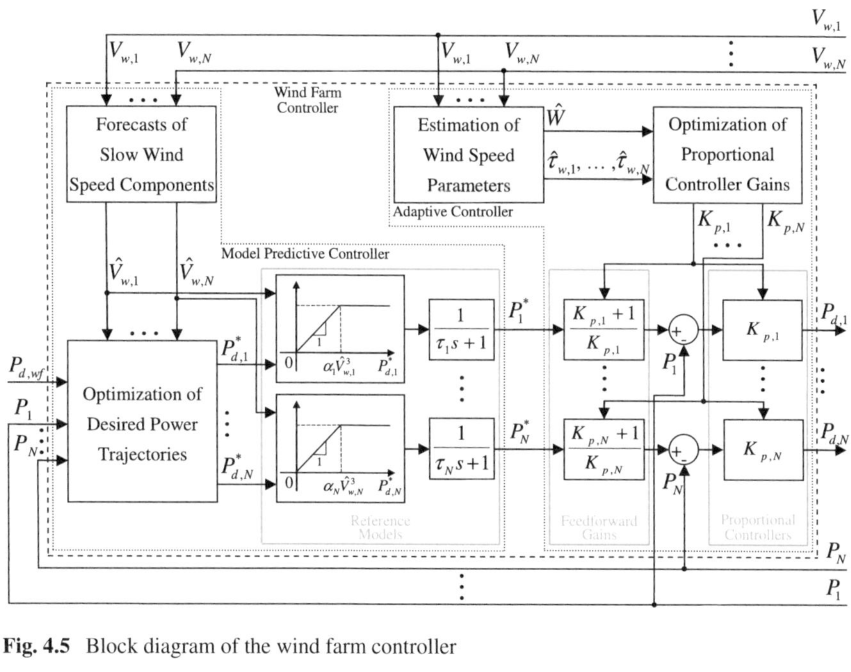

Fig. 4.5 Block diagram of the wind farm controller

$\widehat{W}$ of the covariance matrix W in (4.2), and estimates $\widehat{\tau}_{w,i}$'s of the time constants $\tau_{w,i}$'s in Sect. 4.5.1. The $\widehat{W}$ and $\widehat{\tau}_{w,i}$'s, in turn, are used by the Optimization of Proportional Controller Gains block to calculate the optimal proportional controller gains $K_{p,i}$'s. The $K_{p,i}$'s, along with feedback of the power outputs $P_i(t)$'s and the intermediate power references $P_i^*(t)$'s, are used by the Proportional Controllers and Feedforward Gains blocks to produce the desired power references $P_{d,i}(t)$'s that drive the N WTCSs.

4.4 Model Predictive Controller on Outer Feedback Loop

In this section, we describe the model predictive controller on the outer feedback loop of the WFC, beginning with the Forecasts of Slow Wind Speed Components block and followed by the Optimization of Desired Power Trajectories block. We will not discuss the Reference Models block because it has been fully defined in Fig. 4.5 and has already been discussed in Sect. 4.3.

4.4.1 Forecasts of Slow Wind Speed Components

The current literature offers a large body of work on wind speed forecasting. For instance, [18] develops one of the earliest autoregressive models for simulating and forecasting wind speed on a minute-to-hour timescale, which accounts for several basic characteristics of wind speed such as its positive autocorrelation, non-Gaussian distribution, and diurnal nonstationarity. As another example, [45] from the ANEMOS project and [78] from Argonne National Laboratory each provides an extensive overview of the state-of-the-art in numerical weather prediction models and wind speed/power forecasting models that span different spatial scales (global and regional) and temporal scales (very-short-term, short-term, and medium-term) and that are constructed using different approaches (physical, statistical, and combined). Due to their survey nature, both [45, 78] also contain a large number of references that are devoted to wind speed/power forecasting. As yet another example, [64] from the recently completed Aeolus project [55, 63] uses extended Kalman filters and system identification techniques to develop a forecasting model for effective wind speeds at neighboring turbines in a wind farm, which accounts for atmospheric turbulences and wake effects.

Given that a lot of work has been done on the topic of wind speed forecasting, and given that the focus of this monograph is on wind farm control, we will not delve into this topic here. Rather, we will assume that one of the available models in the literature—such as the forecasting model in [64]—is embedded in the Forecasts of Slow Wind Speed Components block to provide accurate forecasts $\widehat{V}_{w,i}(t)$'s of the slow components $\overline{V}_{w,i}(t)$'s of the wind speeds $V_{w,i}(t)$'s, for as long into the future as is needed by the Optimization of Desired Power Trajectories block.

4.4.2 Optimization of Desired Power Trajectories

In this subsection, we describe the Optimization of Desired Power Trajectories block that is the core of the model predictive controller by introducing a discrete-time predictive model, formulating a tracking accuracy optimization problem, and solving the problem for the optimal desired power trajectories.

4.4.2.1 Discrete-Time Predictive Model

As alluded to in Sect. 4.3.1, at the heart of the model predictive controller is an optimization problem of determining the control inputs $P_{d,i}(t_0)$'s, which needs to be solved at every time $t_0 \geq 0$, which we will refer to as the *current time*. Moreover, this optimization problem must make use of WTCS model (4.4) from the current time t_0 to a future time $t_0 + T$, where $T > 0$ is the optimization horizon. Since (4.4) involves both the fast components $\widetilde{V}_{w,i}(t)$'s and the wind speeds $V_{w,i}(t)$'s, and since their values are not known for all $t \in [t_0, t_0 + T]$, to use (4.4) in the optimization we must replace their values with estimates. To this end, for each $i \in \{1, 2, \ldots, N\}$, note from (4.2) that $\widetilde{V}_{w,i}(t)$ is fast-changing, zero-mean and Gaussian, and from (4.4) that the transfer function from $\widetilde{V}_{w,i}(t)$ to $P_i(t)$ is a low-pass filter $\frac{\tau_i \gamma_i}{\tau_i s+1}$. Therefore, a good estimate of $\widetilde{V}_{w,i}(t)$ for $t \in [t_0, t_0 + T]$ is 0. This, along with (4.1) and the text in Sect. 4.4.1, suggests that a good estimate of $V_{w,i}(t)$ for $t \in [t_0, t_0 + T]$ is the forecast $\widehat{V}_{w,i}(t)$ for $t \in [t_0, t_0 + T]$ provided by the Forecasts of Slow Wind Speed Components block. It follows that we may replace $\widetilde{V}_{w,i}(t)$ and $V_{w,i}(t)$ in (4.4) with 0 and $\widehat{V}_{w,i}(t)$, respectively, to obtain a predictive model

$$\dot{P}_i(t) = -\frac{1}{\tau_i} P_i(t) + \frac{1}{\tau_i} \operatorname{sat}_0^{\alpha_i \widehat{V}_{w,i}^3(t)}(P_{d,i}(t)), \tag{4.6}$$

which is defined at the current time t_0 for $i \in \{1, 2, \ldots, N\}$ and $t \in [t_0, t_0 + T]$. We stress that this predictive model (4.6) is used only by the Optimization of Desired Power Trajectories block (to perform the said optimization) and by the Reference Models block (see Fig. 4.5) of the model predictive controller. The actual WTCS model to be controlled is still (4.4).

To use the predictive model (4.6), which is in continuous-time, for model predictive control that is often implemented in discrete-time, we discretize (4.6) in the following manner: let time $t \geq 0$ be slotted and let each time slot $k \in \{0, 1, 2, \ldots\}$ run from time kT_s to time $(k + 1)T_s$, where $T_s > 0$ is the time slot duration. In addition, let the current time $t_0 \geq 0$ be restricted to the set $\{0, T_s, 2T_s, \ldots\}$ and let $k_0 = t_0/T_s \in \{0, 1, 2, \ldots\}$, so that the current time t_0 is at the beginning of time slot k_0. Moreover, let the optimization horizon $T > 0$ be restricted to the set $\{T_s, 2T_s, 3T_s, \ldots\}$ and let $K = T/T_s \in \{1, 2, 3, \ldots\}$, so that the optimization horizon T occupies K time slots. Furthermore, for simplicity suppose both the control inputs $P_{d,i}(t)$'s and the forecasts $\widehat{V}_{w,i}(t)$'s in (4.6) are constant in each time slot, i.e.,

$P_{d,i}(t) = P_{d,i}[k]$ and $\widehat{V}_{w,i}(t) = \widehat{V}_{w,i}[k]$ for $i \in \{1, 2, \ldots, N\}$, $t \in [kT_s, (k+1)T_s)$, and $k \in \{k_0, k_0+1, \ldots, k_0+K-1\}$. In other words, $P_{d,i}[k]$ and $\widehat{V}_{w,i}[k]$ with square brackets are, respectively, the (constant) values of $P_{d,i}(t)$ and $\widehat{V}_{w,i}(t)$ at time slot k. Then, we may discretize (4.6) as if there is a zero-order hold to obtain a discrete-time predictive model

$$P_i[k+1] = a_i P_i[k] + (1-a_i)P_{d,i}[k], \tag{4.7}$$

which is defined at the current time t_0 for $i \in \{1, 2, \ldots, N\}$ and $k \in \{k_0, k_0 + 1, \ldots, k_0+K-1\}$. In (4.7), $P_i[k]$ with square brackets is the value of the power output $P_i(t)$ at the beginning of time slot k (i.e., $P_i[k] = P_i(kT_s)$), $a_i = e^{-T_s/\tau_i} \in (0, 1)$ is a parameter, and $P_{d,i}[k]$ must satisfy the constraints

$$0 \le P_{d,i}[k] \le \alpha_i \widehat{V}^3_{w,i}[k]. \tag{4.8}$$

Notice that the saturation function appearing in (4.6) does not show up in (4.7) because it is accounted for in (4.8).

To facilitate subsequent use of the discrete-time predictive model (4.7), we rewrite it in a matrix solution form as follows: for each $i \in \{1, 2, \ldots, N\}$, note that (4.7) is a scalar difference equation whose solution is given by

$$\begin{aligned}
P_i[k_0+1] &= a_i P_i[k_0] + (1-a_i)P_{d,i}[k_0],\\
P_i[k_0+2] &= a_i^2 P_i[k_0] + a_i(1-a_i)P_{d,i}[k_0] + (1-a_i)P_{d,i}[k_0+1],\\
&\ \ \vdots\\
P_i[k_0+K] &= a_i^K P_i[k_0] + (1-a_i)\sum_{k=k_0}^{k_0+K-1} a_i^{k_0+K-1-k} P_{d,i}[k],
\end{aligned}$$

which can be expressed in a matrix form as

$$\underbrace{\begin{bmatrix} P_i[k_0+1] \\ P_i[k_0+2] \\ \vdots \\ P_i[k_0+K] \end{bmatrix}}_{\mathbf{P}_i^{k_0+1}} = a_i \underbrace{\begin{bmatrix} 1 \\ a_i \\ \vdots \\ a_i^{K-1} \end{bmatrix}}_{v_i} P_i[k_0] + (1-a_i) \underbrace{\begin{bmatrix} 1 & 0 & \cdots & 0 \\ a_i & 1 & \ddots & \vdots \\ \vdots & \ddots & \ddots & 0 \\ a_i^{K-1} & \cdots & a_i & 1 \end{bmatrix}}_{L_i} \underbrace{\begin{bmatrix} P_{d,i}[k_0] \\ P_{d,i}[k_0+1] \\ \vdots \\ P_{d,i}[k_0+K-1] \end{bmatrix}}_{\mathbf{P}_{d,i}^{k_0}}. \tag{4.9}$$

By letting $\mathbf{P}_i^{k_0+1} \in \mathbb{R}^K$, $v_i \in \mathbb{R}^K$, $L_i \in \mathbb{R}^{K\times K}$, and $\mathbf{P}_{d,i}^{k_0} \in \mathbb{R}^K$ be as defined in (4.9), the equation can be compactly stated as

$$\mathbf{P}_i^{k_0+1} = a_i v_i P_i[k_0] + (1-a_i)L_i \mathbf{P}_{d,i}^{k_0}. \tag{4.10}$$

Furthermore, by letting $\mathbf{P}^{k_0+1} \in \mathbb{R}^{NK}$ and $\mathbf{P}_d^{k_0} \in \mathbb{R}^{NK}$ be defined as

$$\mathbf{P}^{k_0+1} = \begin{bmatrix} \mathbf{P}_1^{k_0+1} \\ \mathbf{P}_2^{k_0+1} \\ \vdots \\ \mathbf{P}_N^{k_0+1} \end{bmatrix}, \quad \mathbf{P}_d^{k_0} = \begin{bmatrix} \mathbf{P}_{d,1}^{k_0} \\ \mathbf{P}_{d,2}^{k_0} \\ \vdots \\ \mathbf{P}_{d,N}^{k_0} \end{bmatrix},$$

and by noting that (4.10) holds for every $i \in \{1, 2, \ldots, N\}$, we obtain

$$\mathbf{P}^{k_0+1} = \begin{bmatrix} a_1 v_1 P_1[k_0] \\ a_2 v_2 P_2[k_0] \\ \vdots \\ a_N v_N P_N[k_0] \end{bmatrix} + \begin{bmatrix} (1-a_1)L_1 & 0 & \cdots & 0 \\ 0 & (1-a_2)L_2 & \ddots & \vdots \\ \vdots & \ddots & \ddots & 0 \\ 0 & \cdots & 0 & (1-a_N)L_N \end{bmatrix} \mathbf{P}_d^{k_0}, \tag{4.11}$$

which is a compact, matrix solution form of (4.7) that will be used shortly.

4.4.2.2 Tracking Accuracy Optimization

With the discrete-time predictive model (4.7) and constraints (4.8) in hand, we next use them to formulate a tracking accuracy optimization problem. To do so, observe that at each current time $t_0 \in \{0, T_s, 2T_s, \ldots\}$ (i.e., at the beginning of each time slot $k_0 \in \{0, 1, 2, \ldots\}$), the Optimization of Desired Power Trajectories block has access to the following information: (i) power outputs $P_i[k_0]$ for $i \in \{1, 2, \ldots, N\}$ from the outer feedback loop; (ii) forecasts $P_{d,wf}[k]$ for $k \in \{k_0, k_0+1, \ldots, k_0+K\}$ from the power grid operator, where $P_{d,wf}[k] = P_{d,wf}(kT_s)$; (iii) control inputs $P_{d,i}[k_0-1]$ for $i \in \{1, 2, \ldots, N\}$ from the previous time slot k_0-1 (when $k_0 = 0$, $P_{d,i}[-1]$ may be defined arbitrarily); and (iv) forecasts $\widehat{V}_{w,i}[k]$ for $i \in \{1, 2, \ldots, N\}$ and $k \in \{k_0, k_0+1, \ldots, k_0+K-1\}$ from the Forecasts of Slow Wind Speed Components block. Thus, at each current time $t_0 \in \{0, T_s, 2T_s, \ldots\}$, this block has all the information it needs to formulate the following optimization problem: find a vector $\mathbf{P}_d^{k_0} \in \mathbb{R}^{NK}$ (i.e., find control inputs $P_{d,i}[k]$ for $i \in \{1, 2, \ldots, N\}$ and $k \in \{k_0, k_0+1, \ldots, k_0+K-1\}$) that minimizes the cost function

$$J_1 = \sum_{k=k_0}^{k_0+K} \eta(k) \left(\sum_{i=1}^{N} P_i[k] - P_{d,wf}[k] \right)^2 \tag{4.12a}$$

$$+ \sum_{k=k_0}^{k_0+K-1} \sum_{i=1}^{N} \mu_i(k) \left(P_{d,i}[k] - P_{d,i}[k-1] \right)^2 \tag{4.12b}$$

$$+\sum_{k=k_0}^{k_0+K-1}\sum_{i=1}^{N}\nu_i(k)P_{d,i}^2[k] \tag{4.12c}$$

subject to the linear dynamics (4.7) and inequality constraints (4.8) for $i \in \{1, 2, \ldots, N\}$ and $k \in \{k_0, k_0+1, \ldots, k_0+K-1\}$, where the $\eta(k)$'s, $\mu_i(k)$'s, and $\nu_i(k)$'s are positive weighting factors. For convenience, we will denote this problem as $\mathcal{P}(k_0)$, emphasizing that it is associated with time slot k_0.

To understand the significance of this constrained, dynamic optimization problem $\mathcal{P}(k_0)$, notice that the optimization variable $\mathbf{P}_d^{k_0}$ is an NK-dimensional vector formed by the N control inputs of the predictive model (4.7) over K time slots into the future. Also note that the cost function J_1 is a sum of three terms labeled as (4.12a)–(4.12c). Since the wind farm power output is given by $P_{wf}(t) = \sum_{i=1}^{N} P_i(t)$, the first term (4.12a) is a receding horizon sum of the squares of the tracking errors, which reflects the tracking accuracy. In contrast, the second term (4.12b) and third term (4.12c) are, respectively, receding horizon sums of the squares of the control input variations and magnitudes, which reflect the control effort. Note that penalizing the control input variations helps prevent dramatic jumps in the turbine states that may cause undesirable mechanical vibrations. Lastly, note that the weighting factors $\eta(k)$'s, $\mu_i(k)$'s, and $\nu_i(k)$'s reflect the relative importance of the summands for different wind turbines at different time slots. Hence, solving the optimization problem $\mathcal{P}(k_0)$ for the optimal $\mathbf{P}_d^{k_0}$ may be regarded as *finding the desired power trajectories that optimize a weighted combination of the tracking accuracy and control effort over a finite horizon.*

Although $\mathcal{P}(k_0)$ is a dynamic optimization problem, it may be converted into a static one as follows: observe that the first term (4.12a) of the cost function J_1 can be written as

$$\begin{aligned}
&\sum_{k=k_0}^{k_0+K}\eta(k)\left(\sum_{i=1}^{N}P_i[k]-P_{d,wf}[k]\right)^2\\
&=\sum_{k=k_0+1}^{k_0+K}\eta(k)\left(\sum_{i=1}^{N}P_i[k]-P_{d,wf}[k]\right)^2+\eta(k_0)\left(\sum_{i=1}^{N}P_i[k_0]-P_{d,wf}[k_0]\right)^2\\
&=\sum_{k=k_0+1}^{k_0+K}\eta(k)\left(\sum_{i=1}^{N}P_i[k]\right)^2-2\sum_{k=k_0+1}^{k_0+K}\eta(k)\sum_{i=1}^{N}P_i[k]P_{d,wf}[k]\\
&\quad+\sum_{k=k_0+1}^{k_0+K}\eta(k)P_{d,wf}^2[k]+\eta(k_0)\left(\sum_{i=1}^{N}P_i[k_0]-P_{d,wf}[k_0]\right)^2\\
&=(\mathbf{P}^{k_0+1})^T\begin{bmatrix}H^{k_0+1} & H^{k_0+1} & \cdots & H^{k_0+1}\\ H^{k_0+1} & H^{k_0+1} & \cdots & H^{k_0+1}\\ \vdots & \vdots & \ddots & \vdots\\ H^{k_0+1} & H^{k_0+1} & \cdots & H^{k_0+1}\end{bmatrix}\mathbf{P}^{k_0+1}
\end{aligned}$$

$$
-2\begin{bmatrix}\mathbf{P}_{d,wf}^{k_0+1}\\ \mathbf{P}_{d,wf}^{k_0+1}\\ \vdots\\ \mathbf{P}_{d,wf}^{k_0+1}\end{bmatrix}^T\begin{bmatrix}H^{k_0+1} & 0 & \cdots & 0\\ 0 & H^{k_0+1} & \ddots & \vdots\\ \vdots & \ddots & \ddots & 0\\ 0 & \cdots & 0 & H^{k_0+1}\end{bmatrix}\mathbf{P}^{k_0+1}
+\sum_{k=k_0+1}^{k_0+K}\eta(k)P_{d,wf}^2[k]+\eta(k_0)\left(\sum_{i=1}^{N}P_i[k_0]-P_{d,wf}[k_0]\right)^2, \tag{4.13}
$$

where $H^{k_0+1}\in\mathbb{R}^{K\times K}$ and $\mathbf{P}_{d,wf}^{k_0+1}\in\mathbb{R}^K$ are defined as

$$
H^{k_0+1}=\begin{bmatrix}\eta(k_0+1) & 0 & \cdots & 0\\ 0 & \eta(k_0+2) & \ddots & \vdots\\ \vdots & \ddots & \ddots & 0\\ 0 & \cdots & 0 & \eta(k_0+K)\end{bmatrix},\quad \mathbf{P}_{d,wf}^{k_0+1}=\begin{bmatrix}P_{d,wf}[k_0+1]\\ P_{d,wf}[k_0+2]\\ \vdots\\ P_{d,wf}[k_0+K]\end{bmatrix}.
$$

Substituting (4.11) into (4.13), the first term (4.12a) becomes

$$
\begin{aligned}
&\sum_{k=k_0}^{k_0+K}\eta(k)\left(\sum_{i=1}^{N}P_i[k]-P_{d,wf}[k]\right)^2\\
&=(\mathbf{P}_d^{k_0})^T\begin{bmatrix}(1-a_1)(1-a_1)L_1^TH^{k_0+1}L_1 & \cdots & (1-a_1)(1-a_N)L_1^TH^{k_0+1}L_N\\ \vdots & \ddots & \vdots\\ (1-a_N)(1-a_1)L_N^TH^{k_0+1}L_1 & \cdots & (1-a_N)(1-a_N)L_N^TH^{k_0+1}L_N\end{bmatrix}\mathbf{P}_d^{k_0}\\
&+2\begin{bmatrix}a_1v_1P_1[k_0]\\ \vdots\\ a_Nv_NP_N[k_0]\end{bmatrix}^T\begin{bmatrix}(1-a_1)H^{k_0+1}L_1 & \cdots & (1-a_N)H^{k_0+1}L_N\\ \vdots & \ddots & \vdots\\ (1-a_1)H^{k_0+1}L_1 & \cdots & (1-a_N)H^{k_0+1}L_N\end{bmatrix}\mathbf{P}_d^{k_0}\\
&-2\begin{bmatrix}\mathbf{P}_{d,wf}^{k_0+1}\\ \vdots\\ \mathbf{P}_{d,wf}^{k_0+1}\end{bmatrix}^T\begin{bmatrix}(1-a_1)H^{k_0+1}L_1 & & 0\\ & \ddots & \\ 0 & & (1-a_N)H^{k_0+1}L_N\end{bmatrix}\mathbf{P}_d^{k_0}\\
&-2\sum_{i=1}^{N}a_iP_i[k_0](\mathbf{P}_{d,wf}^{k_0+1})^TH^{k_0+1}v_i+\sum_{i=1}^{N}\sum_{j=1}^{N}a_ia_jP_i[k_0]P_j[k_0]v_i^TH^{k_0+1}v_j\\
&+\sum_{k=k_0+1}^{k_0+K}\eta(k)P_{d,wf}^2[k]+\eta(k_0)\left(\sum_{i=1}^{N}P_i[k_0]-P_{d,wf}[k_0]\right)^2.
\end{aligned}\tag{4.14}
$$

Similarly, observe that the second term (4.12b) and third term (4.12c) of the cost function J_1 can be expressed as

$$
\begin{aligned}
&\sum_{k=k_0}^{k_0+K-1}\sum_{i=1}^{N}\mu_i(k)\left(P_{d,i}[k]-P_{d,i}[k-1]\right)^2+\sum_{k=k_0}^{k_0+K-1}\sum_{i=1}^{N}\nu_i(k)P_{d,i}^2[k]\\
&=(\mathbf{P}_d^{k_0})^T\begin{bmatrix} U_1^{k_0}+V_1^{k_0} & 0 & \cdots & 0\\ 0 & U_2^{k_0}+V_2^{k_0} & \ddots & \vdots\\ \vdots & \ddots & \ddots & 0\\ 0 & \cdots & 0 & U_N^{k_0}+V_N^{k_0}\end{bmatrix}\mathbf{P}_d^{k_0}\\
&-2\left(\begin{bmatrix}\mu_1(k_0)P_{d,1}[k_0-1]\\ \mu_2(k_0)P_{d,2}[k_0-1]\\ \vdots\\ \mu_N(k_0)P_{d,N}[k_0-1]\end{bmatrix}\otimes\begin{bmatrix}1\\0\\ \vdots\\0\end{bmatrix}\right)^T\mathbf{P}_d^{k_0}+\sum_{i=1}^{N}\mu_i(k_0)P_{d,i}^2[k_0-1],
\end{aligned}
\tag{4.15}
$$

where $U_i^{k_0}\in\mathbb{R}^{K\times K}$ and $V_i^{k_0}\in\mathbb{R}^{K\times K}$ are defined for $i\in\{1,2,\ldots,N\}$ as

$$
U_i^{k_0}=\begin{bmatrix}
\widetilde{\mu}_i^0+\widetilde{\mu}_i^1 & -\widetilde{\mu}_i^1 & 0 & 0 & \cdots & 0\\
-\widetilde{\mu}_i^1 & \widetilde{\mu}_i^1+\widetilde{\mu}_i^2 & -\widetilde{\mu}_i^2 & 0 & & 0\\
0 & -\widetilde{\mu}_i^2 & \widetilde{\mu}_i^2+\widetilde{\mu}_i^3 & \ddots & \ddots & \vdots\\
0 & 0 & \ddots & \ddots & -\widetilde{\mu}_i^{K-2} & 0\\
\vdots & & \ddots & -\widetilde{\mu}_i^{K-2} & \widetilde{\mu}_i^{K-2}+\widetilde{\mu}_i^{K-1} & -\widetilde{\mu}_i^{K-1}\\
0 & 0 & \cdots & 0 & -\widetilde{\mu}_i^{K-1} & \widetilde{\mu}_i^{K-1}
\end{bmatrix},
$$

$$
V_i^{k_0}=\begin{bmatrix}
\nu_i(k_0) & 0 & \cdots & 0\\
0 & \nu_i(k_0+1) & \ddots & \vdots\\
\vdots & \ddots & \ddots & 0\\
0 & \cdots & 0 & \nu_i(k_0+K-1)
\end{bmatrix},
$$

$\otimes$ denotes the Kronecker product, and $\widetilde{\mu}_i^m$ is a shorthand for $\mu_i(k_0+m)$. Examining (4.14) and (4.15), we see that the cost function J_1 in (4.12) is a quadratic function of the optimization variable $\mathbf{P}_d^{k_0}$, i.e.,

$$
J_1=(\mathbf{P}_d^{k_0})^T S^{k_0}\mathbf{P}_d^{k_0}+(b^{k_0})^T\mathbf{P}_d^{k_0}+c^{k_0} \tag{4.16}
$$

for some $S^{k_0}\in\mathbb{R}^{NK\times NK}$, $b^{k_0}\in\mathbb{R}^{NK}$, and $c^{k_0}\in\mathbb{R}$, which can be determined from (4.14) and (4.15). In fact, as Proposition 4.1 below shows, S^{k_0} is a symmetric positive definite matrix. Therefore, by using (4.11), the dynamic optimization problem $\mathcal{P}(k_0)$ may be converted into a static, convex optimization problem of finding $\mathbf{P}_d^{k_0}\in\mathbb{R}^{NK}$ that minimizes the (convex) quadratic cost function (4.16) subject to the rectangular constraints (4.8). Obviously, this problem is a quadratic program, which we will denote as $\mathcal{P}'(k_0)$.

Proposition 4.1 *The matrix S^{k_0} in* (4.16) *is symmetric positive definite.*

Proof Observe from (4.14) and (4.15) that

$$S^{k_0} = \begin{bmatrix} (1-a_1)L_1^T & & 0 \\ & \ddots & \\ 0 & & (1-a_N)L_N^T \end{bmatrix} \left(\mathbb{1}_N \mathbb{1}_N^T \otimes H^{k_0+1}\right) \begin{bmatrix} (1-a_1)L_1 & & 0 \\ & \ddots & \\ 0 & & (1-a_N)L_N \end{bmatrix}$$
$$+ \begin{bmatrix} U_1^{k_0}+V_1^{k_0} & & 0 \\ & \ddots & \\ 0 & & U_N^{k_0}+V_N^{k_0} \end{bmatrix}, \tag{4.17}$$

where $\mathbb{1}_N \in \mathbb{R}^N$ is the all-one column vector. Clearly, S^{k_0} is symmetric. Since the $\eta(k)$'s are positive, H^{k_0+1} is positive definite. Thus, $\mathbb{1}_N \mathbb{1}_N^T \otimes H^{k_0+1}$ is positive semidefinite, and so is the first term in (4.17). Since the $\mu_i(k)$'s are positive, without the constant $\widetilde{\mu}_i^0$ in its $(1, 1)$ entry, $U_i^{k_0}$ would be a weighted Laplacian matrix of an undirected path graph, which is positive semidefinite with exactly one eigenvalue at 0, whose corresponding eigenvector is $\mathbb{1}_K$. Hence, with the constant $\widetilde{\mu}_i^0$ in its $(1, 1)$ entry, $U_i^{k_0}$ would be positive definite. Finally, since the $\nu_i(k)$'s are positive, $V_i^{k_0}$ is also positive definite. Therefore, the second term in (4.17) is positive definite, and so is S^{k_0}. □

4.4.2.3 Optimal Desired Power Trajectories

Because its cost function (4.16) is strongly convex and its feasible set (4.8) is nonempty, convex, and closed, the optimization problem $\mathcal{P}'(k_0)$ always has a unique solution $\mathbf{P}_d^{k_0} \in \mathbb{R}^{NK}$. Unfortunately however, the solution $\mathbf{P}_d^{k_0}$ cannot, in general, be analytically obtained [16]. Nevertheless, effective and reliable numerical algorithms, capable of solving the optimization problem $\mathcal{P}'(k_0)$ in a few tens of seconds when its dimension NK is up to several thousands, are currently available (e.g., the interior-point methods [16] and augmented Lagrangian methods [13]). Indeed, a number of these algorithms have been implemented in free and commercially available software packages (e.g., AMPL [53], CVXOPT [54], MATLAB, and OOQP [44], among many others). Hence, in what follows we will assume that one of the available algorithms is embedded in the Optimization of Desired Power Trajectories block.

With such an algorithm, the Optimization of Desired Power Trajectories block can implement model predictive control by taking the following actions at each current time $t_0 \in \{0, T_s, 2T_s, \ldots\}$ (i.e., at the beginning of each time slot $k_0 \in \{0, 1, 2, \ldots\}$): first, the block uses the information it has (i.e., power outputs $P_i[k_0]$ for $i \in \{1, 2, \ldots, N\}$, forecasts $P_{d,wf}[k]$ for $k \in \{k_0, k_0+1, \ldots, k_0+K\}$, control inputs $P_{d,i}[k_0-1]$ for $i \in \{1, 2, \ldots, N\}$, and forecasts $\widehat{V}_{w,i}[k]$ for $i \in \{1, 2, \ldots, N\}$ and $k \in \{k_0, k_0+1, \ldots, k_0+K-1\}$) to formulate the optimization problem $\mathcal{P}'(k_0)$ (i.e., calculate S^{k_0} and b^{k_0} in (4.16) and $\alpha_i \widehat{V}_{w,i}^3[k]$ in (4.8)). Next, the block uses the algorithm embedded in it to solve $\mathcal{P}'(k_0)$ for the optimal desired power trajectories

$\mathbf{P}_d^{k_0}$ (i.e., $P_{d,i}[k]$ for $i \in \{1, 2, \ldots, N\}$ and $k \in \{k_0, k_0 + 1, \ldots, k_0 + K - 1\}$). Subsequently, the block uses a zero-order hold to convert the N real numbers $P_{d,i}[k_0]$ for $i \in \{1, 2, \ldots, N\}$ (i.e., the initial portion of $\mathbf{P}_d^{k_0}$) into N continuous-time signals denoted as $P_{d,i}^*(t)$ for $i \in \{1, 2, \ldots, N\}$. Lastly, the block lets these signals be its outputs during time slot k_0, which runs from time t_0 to time $t_0 + T_s$.

Notice from the aforementioned actions that upon solving the optimization problem at time t_0 (which incorporates the feedback at time t_0 and forecasts from time t_0 to time t_0+T), the Optimization of Desired Power Trajectories block applies only the *first step* of the solution from time t_0 to time $t_0 + T_s$, and moves "immediately" to the next optimization problem at time $t_0 + T_s$. This idea of model predictive control allows the block to better utilize not only the available feedback (on the power outputs), but also the latest forecasts (on the desired wind farm power output and slow wind speed components), which tend to be more accurate than outdated ones. Therefore, the block produces optimal desired power trajectories $P_{d,i}^*(t)$'s that account for what is happening (i.e., feedback), and what will likely happen (i.e., forecasts).

4.5 Adaptive Controller on Inner Feedback Loop

In this section, we describe the adaptive controller on the inner feedback loop of the WFC, beginning with the Estimation of Wind Speed Parameters block, followed by the Proportional Controllers and Feedforward Gains blocks, and ending with the Optimization of Proportional Controller Gains block.

4.5.1 *Estimation of Wind Speed Parameters*

As pointed out in Sect. 4.3.1, the model predictive controller is not intended to make $P_{wf}(t)$ smooth on a shorter timescale by rejecting the fast-changing $\widetilde{V}_{w,i}(t)$'s in (4.4). Such a goal is the responsibility of the adaptive controller, which we will now develop. To begin, recall from (4.2) that the fast wind speed components $\widetilde{V}_{w,1}(t), \ldots, \widetilde{V}_{w,N}(t) \in \mathbb{R}$ are modeled as the output of a general, N-by-N_w asymptotically stable transfer function matrix $G_w(s)$, whose input is a stationary, zero-mean white Gaussian random process $w(t) \in \mathbb{R}^{N_w}$ with covariance matrix $W \in \mathbb{R}^{N_w \times N_w}$. Although we could proceed with such a general $G_w(s)$, dealing with a specific one greatly simplifies the development. Thus, in what follows we will assume that $G_w(s)$ is a square (i.e., $N_w = N$), diagonal, first-order low-pass filter of the form

$$G_w(s) = \begin{bmatrix} \frac{1}{\tau_{w,1}s+1} & 0 & \cdots & 0 \\ 0 & \frac{1}{\tau_{w,2}s+1} & \ddots & \vdots \\ \vdots & \ddots & \ddots & 0 \\ 0 & \cdots & 0 & \frac{1}{\tau_{w,N}s+1} \end{bmatrix}, \tag{4.18}$$

where $\tau_{w,i} > 0$ is the time constant. We note that other choices of $G_w(s)$ are possible and can be treated in the same way as (4.18) in the sequel. With (4.2) and (4.18), the fast wind speed components $\widetilde{V}_{w,i}(t)$'s are described by

$$\dot{\widetilde{V}}_{w,i}(t) = -\frac{1}{\tau_{w,i}}\widetilde{V}_{w,i}(t) + \frac{1}{\tau_{w,i}}w_i(t), \tag{4.19}$$

implying that they are stationary, zero-mean colored Gaussian random processes that are completely specified by the covariance matrix W and time constants $\tau_{w,i}$'s. Since the values of these wind speed parameters are not readily known, and since these values generally change slowly over time due to gradual changes in wind speed characteristics, we will assume that one of the available parameter estimation techniques in the literature—such as those techniques in [71]—is embedded in the Estimation of Wind Speed Parameters block to periodically provide an accurate estimate $\widehat{W}$ of the covariance matrix W, and accurate estimates $\widehat{\tau}_{w,i}$'s of the time constants $\tau_{w,i}$'s, based on noisy measurements of the past and present wind speeds $V_{w,i}(t)$'s. These estimates will be used at the end of Sect. 4.5.3.3.

4.5.2 *Proportional Controllers and Feedforward Gains*

Another major component of the adaptive controller, which we will now describe, is the Proportional Controllers and Feedforward Gains blocks. To appreciate the role of these blocks, observe that the fast components $\widetilde{V}_{w,i}(t)$'s appear as "disturbances" in (4.4), the *actual* WTCS model to be controlled. However, these $\widetilde{V}_{w,i}(t)$'s do not appear in (4.6), the *predictive* model used by the Optimization of Desired Power Trajectories block to compute the staircase-like, optimal desired power trajectories $P^*_{d,i}(t)$'s. Hence, if these $P^*_{d,i}(t)$'s were the control inputs $P_{d,i}(t)$'s in (4.4)—that is, if the signals $P^*_{d,1}, \ldots, P^*_{d,N}$ on the left of Fig. 4.5 were to bypass the Reference Models, Feedforward Gains, and Proportional Controllers blocks and become the signals $P_{d,1}, \ldots, P_{d,N}$ on the right—they would not be able to reject the disturbances $\widetilde{V}_{w,i}(t)$'s. As a result, the power outputs $P_i(t)$'s in (4.4) might fluctuate substantially, causing the wind farm power output $P_{wf}(t) = \sum_{i=1}^{N} P_i(t)$ to be choppy and not smooth. To reject the $\widetilde{V}_{w,i}(t)$'s, we first introduce the Reference Models block, which is essentially a disturbance-less copy of (4.4), to convert the staircase-like $P^*_{d,i}(t)$'s into slowly-varying, intermediate power references $P^*_i(t)$'s, which we would like the $P_i(t)$'s to track. We then introduce a set of N decoupled proportional controllers that, for each $i \in \{1, 2, \ldots, N\}$, let the control input $P_{d,i}(t)$ be directly proportional to the tracking error $P^*_i(t) - P_i(t)$, i.e.,

$$P_{d,i}(t) = K_{p,i}(P^*_i(t) - P_i(t)), \tag{4.20}$$

where $K_{p,i} > 0$ is the proportional controller gain. Because (4.4) is a type 0 plant in the linear region, to ensure a zero steady-state error when $P^*_i(t)$ is constant and

$\widetilde{V}_{w,i}(t)$ is absent, we further precede each proportional controller with a feedforward gain of $\frac{1+K_{p,i}}{K_{p,i}}$, so that control law (4.20) becomes

$$P_{d,i}(t) = K_{p,i}\left(\frac{1+K_{p,i}}{K_{p,i}}P_i^*(t) - P_i(t)\right). \tag{4.21}$$

Notice that control law (4.21) is represented inside the Proportional Controllers and Feedforward Gains blocks in Fig. 4.5. Also note that this control law is chosen mainly for simplicity, and other more sophisticated choices (e.g., N decoupled PID controllers or a linear quadratic regulator) are possible at the expense of a higher system order, a more complicated stability and performance analysis, and a larger number of controller parameters to tune.

4.5.3 *Optimization of Proportional Controller Gains*

In this subsection, we describe the Optimization of Proportional Controller Gains block that is the core of the adaptive controller by analyzing the inner feedback loop, formulating a smoothness optimization problem, and solving the problem for the optimal proportional controller gains.

4.5.3.1 Inner Feedback Loop

Having specified the control law (4.21), we now analyze the inner feedback loop described by (4.4), (4.19), and (4.21). To facilitate the analysis, we assume that the slowly-varying, intermediate power references $P_i^*(t)$'s in (4.21) are so slow that they may be treated as constants, i.e., $P_i^*(t) = P_i^*$. In addition, as is often done in linear control systems analysis, we assume that the saturation function in (4.4) is not present, so that (4.4) is completely linear. With these assumptions and the substitution of (4.21) into (4.4), we obtain

$$\dot{P}_i(t) = -\frac{1+K_{p,i}}{\tau_i}P_i(t) + \frac{1+K_{p,i}}{\tau_i}P_i^* + \gamma_i \widetilde{V}_{w,i}(t), \tag{4.22}$$

which is valid for every $i \in \{1, 2, \ldots, N\}$. Observe that when there is no disturbance, i.e., $\widetilde{V}_{w,i}(t) = 0$, system (4.22) has a unique equilibrium point, denoted as $P_{i,eq}$, that is given by

$$P_{i,eq} = P_i^*, \tag{4.23}$$

which is ideal since P_i^* is the amount of power the model predictive controller wants WTCS i to generate. Moreover, according to (4.21), the control input $P_{d,i}(t)$

evaluated at this equilibrium point, denoted as $P_{d,i,eq}$, is given by

$$P_{d,i,eq} = P_i^*. \tag{4.24}$$

To translate the equilibrium point $P_{i,eq}$ of system (4.22) to the origin, and to redefine the control input $P_{d,i}(t)$ in (4.21) so that it is zero at $P_{d,i,eq}$, let

$$\Delta P_i(t) = P_i(t) - P_{i,eq}, \tag{4.25}$$

$$\Delta P_{d,i}(t) = P_{d,i}(t) - P_{d,i,eq}. \tag{4.26}$$

Then, the translated system with new state variable $\Delta P_i(t)$ can be written using (4.22), (4.23), and (4.25) as

$$\Delta \dot{P}_i(t) = -\frac{1 + K_{p,i}}{\tau_i}\Delta P_i(t) + \gamma_i \widetilde{V}_{w,i}(t), \tag{4.27}$$

and the new control input $\Delta P_{d,i}(t)$ can be written using (4.21) and (4.23)–(4.26) as

$$\Delta P_{d,i}(t) = -K_{p,i}\Delta P_i(t). \tag{4.28}$$

By merging the dynamics of $\widetilde{V}_{w,i}(t)$ and $\Delta P_i(t)$ in (4.19) and (4.27), we get

$$\begin{bmatrix} \dot{\widetilde{V}}_{w,i}(t) \\ \Delta \dot{P}_i(t) \end{bmatrix} = \begin{bmatrix} -\frac{1}{\tau_{w,i}} & 0 \\ \gamma_i & -\frac{1+K_{p,i}}{\tau_i} \end{bmatrix} \begin{bmatrix} \widetilde{V}_{w,i}(t) \\ \Delta P_i(t) \end{bmatrix} + \begin{bmatrix} \frac{1}{\tau_{w,i}} \\ 0 \end{bmatrix} w_i(t). \tag{4.29}$$

Since (4.29) holds for every $i \in \{1, 2, \ldots, N\}$, we can rewrite it as

$$\begin{bmatrix} \dot{\widetilde{V}}_{w}(t) \\ \Delta \dot{P}(t) \end{bmatrix} = \underbrace{\begin{bmatrix} A_{11} & 0 \\ A_{21} & A_{22} \end{bmatrix}}_{A} \begin{bmatrix} \widetilde{V}_{w}(t) \\ \Delta P(t) \end{bmatrix} + \underbrace{\begin{bmatrix} -A_{11} \\ 0 \end{bmatrix}}_{B} w(t), \tag{4.30}$$

where $A \in \mathbb{R}^{2N\times 2N}$ and $B \in \mathbb{R}^{2N\times N}$ are as defined in (4.30), and

$$\begin{aligned}
\widetilde{V}_w(t) &= \left[\widetilde{V}_{w,1}(t)\ \widetilde{V}_{w,2}(t)\ \cdots\ \widetilde{V}_{w,N}(t)\right]^T, \\
\Delta P(t) &= \left[\Delta P_1(t)\ \Delta P_2(t)\ \cdots\ \Delta P_N(t)\right]^T, \\
A_{11} &= \operatorname{diag}\left(-\frac{1}{\tau_{w,1}}, -\frac{1}{\tau_{w,2}}, \ldots, -\frac{1}{\tau_{w,N}}\right), \\
A_{21} &= \operatorname{diag}(\gamma_1, \gamma_2, \ldots, \gamma_N), \\
A_{22} &= \operatorname{diag}\left(-\frac{1 + K_{p,1}}{\tau_1}, -\frac{1 + K_{p,2}}{\tau_2}, \ldots, -\frac{1 + K_{p,N}}{\tau_N}\right).
\end{aligned}$$

Expression (4.30) describes the dynamics on the inner feedback loop. Proposition 4.2 below shows that these dynamics have an appealing feature:

Proposition 4.2 *For any* $\tau_{w,1}, \tau_{w,2}, \ldots, \tau_{w,N} > 0$, *any* $\tau_1, \tau_2, \ldots, \tau_N > 0$, *any* $\gamma_1, \gamma_2, \ldots, \gamma_N \geq 0$, *and any* $K_{p,1}, K_{p,2}, \ldots, K_{p,N} > 0$, *the matrix* A *in* (4.30) *is asymptotically stable. Furthermore, system* (4.30) *is bounded-input bounded-state (BIBS) stable.*

Proof Notice that A in (4.30) is a lower triangular matrix whose diagonal entries are all negative. Therefore, A is asymptotically stable. It follows that system (4.30) is BIBS stable. □

4.5.3.2 Smoothness Optimization

With the inner feedback loop dynamics (4.30) in hand, we next use it to formulate a smoothness optimization problem. To do so, observe that the Optimization of Proportional Controller Gains block has access to the following information: (i) an accurate estimate $\widehat{W}$ of the covariance matrix W and accurate estimates $\widehat{\tau}_{w,i}$'s of the time constants $\tau_{w,i}$'s from the Estimation of Wind Speed Parameters block; and (ii) the time constants τ_i's and scalar gains γ_i's as allowed by the problem statement in Sect. 4.2.3. Thus, in view of (4.30) and (4.28), this block has all the information it needs to formulate the following optimization problem: find proportional controller gains $K_{p,i} > 0$ for $i \in \{1, 2, \ldots, N\}$ that minimize the cost function

$$J_2 = \lim_{t\to\infty} \mathrm{E}\left\{\left(\sum_{i=1}^{N} \Delta P_i(t)\right)^2 + \sum_{i=1}^{N} \epsilon_i \Delta P_{d,i}^2(t)\right\}, \tag{4.31}$$

where the ϵ_i's are positive weighting factors.

To understand why this stochastic optimization problem is of interest, note that its cost function J_2 in (4.31) is a sum of two terms, namely,

$$\lim_{t\to\infty} \mathrm{E}\left\{\left(\sum_{i=1}^{N} \Delta P_i(t)\right)^2\right\} \quad \text{and} \quad \sum_{i=1}^{N} \epsilon_i \lim_{t\to\infty} \mathrm{E}\{\Delta P_{d,i}^2(t)\}.$$

Because $\sum_{i=1}^{N} \Delta P_i(t) = P_{wf}(t) - \sum_{i=1}^{N} P_i^*$ according to (4.25), (4.23), and the definition of wind farm power output $P_{wf}(t)$, and because $\sum_{i=1}^{N} P_i^*$ is the ideal amount of power the model predictive controller wants the wind farm to generate, the first term is the steady-state variance of the regulation error, which reflects the smoothness of $P_{wf}(t)$. In contrast, because of (4.26) and (4.24), the second term is a weighted sum of the steady-state variances of the control inputs, which reflects the control effort. Hence, solving the stochastic optimization problem for the optimal $K_{p,i}$'s may be regarded as *finding the proportional controller gains that optimize* *a weighted combination of the smoothness and control effort in steady-state.* In particular, small ϵ_i's correspond to a cheap control design where smoothness is emphasized more, whereas large ϵ_i's correspond to an expensive one where control effort is.

4.5.3.3 Optimal Proportional Controller Gains

To solve the aforementioned optimization problem, we first show that its cost function J_2 defined in (4.31) can be expressed in a tractable form. To do that, consider the following standard result in stochastic linear systems theory [67]:

Lemma 4.1 *Consider a continuous-time linear time-invariant system*

$$\dot{\mathsf{x}}(t) = \mathsf{A}\mathsf{x}(t) + \mathsf{w}(t),$$

where $\mathsf{x}(t) \in \mathbb{R}^{\mathsf{n}}$ *is its state,* $\mathsf{A} \in \mathbb{R}^{\mathsf{n}\times\mathsf{n}}$ *is its system matrix, and* $\mathsf{w}(t) \in \mathbb{R}^{\mathsf{n}}$ *is a stationary, zero-mean white Gaussian random process with a symmetric positive semidefinite covariance matrix* $\mathsf{W} \in \mathbb{R}^{\mathsf{n}\times\mathsf{n}}$. *If* A *is asymptotically stable, then for any symmetric matrix* $\mathsf{Q} \in \mathbb{R}^{\mathsf{n}\times\mathsf{n}}$,

$$\lim_{t\to\infty} \mathrm{E}\{\mathsf{x}^T(t)\mathsf{Q}\mathsf{x}(t)\} = \mathrm{trace}(\mathsf{S}\mathsf{Q}),$$

where $\mathsf{S} \in \mathbb{R}^{\mathsf{n}\times\mathsf{n}}$ *is the unique, symmetric positive semidefinite solution to the Lyapunov equation*

$$\mathsf{A}\mathsf{S} + \mathsf{S}\mathsf{A}^T = -\mathsf{W}.$$

Notice that the context of Lemma 4.1 matches the one of system (4.30). Indeed, the n, $\mathsf{x}(t)$, A, $\mathsf{w}(t)$, and W in the lemma may be thought of as the $2N$, $[\widetilde{V}_w^T(t)\ \Delta P^T(t)]^T$, A, $Bw(t)$, and BWB^T in (4.30), respectively, where the latter is due to the fact that $\mathsf{W}\delta(t-\tau) = \mathrm{E}\{\mathsf{w}(t)\mathsf{w}^T(\tau)\} = \mathrm{E}\{Bw(t)w^T(\tau)B^T\} = BWB^T\delta(t-\tau)$. Therefore, the lemma is applicable if: (i) A in (4.30) is asymptotically stable; and (ii) J_2 in (4.31) can be cast into the form of $\lim_{t\to\infty} \mathrm{E}\{\mathsf{x}^T(t)\mathsf{Q}\mathsf{x}(t)\}$ for some Q. To establish (i), note from the text below (4.18), (4.3), and (4.20) that $\tau_{w,i} > 0$, $\tau_i > 0$, $\gamma_i \geq 0$, and $K_{p,i} > 0$ for every $i \in \{1, 2, \ldots, N\}$. Thus, by Proposition 4.2, (i) is satisfied. To establish (ii), observe from (4.28) and (4.31) that

$$\begin{aligned} J_2 &= \lim_{t\to\infty} \mathrm{E}\left\{\left(\sum_{i=1}^{N} \Delta P_i(t)\right)^2 + \sum_{i=1}^{N} \epsilon_i K_{p,i}^2 \Delta P_i^2(t)\right\} \\ &= \lim_{t\to\infty} \mathrm{E}\left\{ \left[\widetilde{V}_w^T(t)\ \Delta P^T(t)\right] \underbrace{\begin{bmatrix} 0 & 0 \\ 0 & Q_{22} \end{bmatrix}}_{Q} \begin{bmatrix} \widetilde{V}_w(t) \\ \Delta P(t) \end{bmatrix} \right\}, \end{aligned} \tag{4.32}$$

where $Q \in \mathbb{R}^{2N\times 2N}$ is as defined in (4.32) and

$$Q_{22} = \begin{bmatrix} 1+\epsilon_1 K_{p,1}^2 & 1 & \cdots & 1 \\ 1 & 1+\epsilon_2 K_{p,2}^2 & \ddots & \vdots \\ \vdots & \ddots & \ddots & 1 \\ 1 & \cdots & 1 & 1+\epsilon_N K_{p,N}^2 \end{bmatrix}.$$

Hence, (ii) is met with $\mathsf{Q} = Q$. It follows from Lemma 4.1 and (4.32) that

$$J_2 = \text{trace}(SQ), \tag{4.33}$$

where $S \in \mathbb{R}^{2N\times 2N}$ is the unique, symmetric positive semidefinite solution to the Lyapunov equation

$$AS + SA^T = -BWB^T. \tag{4.34}$$

Having expressed the cost function J_2 defined in (4.31) in a tractable form (4.33) and (4.34), we next show that it can be written as an explicit function of the relevant parameters. To accomplish that, let S be partitioned as

$$S = \begin{bmatrix} S_1 & S_2 \\ S_2^T & S_3 \end{bmatrix}, \tag{4.35}$$

where $S_1 = S_1^T \in \mathbb{R}^{N\times N}$, $S_2 \in \mathbb{R}^{N\times N}$, and $S_3 = S_3^T \in \mathbb{R}^{N\times N}$. Then, because of how A and B are partitioned in (4.30), Eq. (4.34) can be written as

$$A_{11}S_1 + S_1 A_{11}^T = -A_{11} W A_{11}^T, \tag{4.36}$$

$$A_{11}S_2 + S_2 A_{22}^T = -S_1 A_{21}^T, \tag{4.37}$$

$$A_{22}S_3 + S_3 A_{22}^T = -A_{21} S_2 - S_2^T A_{21}^T. \tag{4.38}$$

To solve (4.36)–(4.38), notice from (4.30) that A_{11}, A_{21}, and A_{22} are diagonal matrices with A_{11} and A_{22} being asymptotically stable. Therefore, the Lyapunov equation (4.36), Sylvester equation (4.37), and Lyapunov equation (4.38) each has a unique solution that can be easily determined. Indeed, by denoting the (i, j) entry of S_1, S_2, S_3, and W as $S_{1,ij}$, $S_{2,ij}$, $S_{3,ij}$, and W_{ij}, respectively, for every $i, j \in \{1, 2, \ldots, N\}$, the solutions may be stated as

$$S_{1,ij} = \frac{W_{ij}}{\tau_{w,i} + \tau_{w,j}}, \tag{4.39}$$

$$S_{2,ij} = \frac{\gamma_j S_{1,ij}}{\frac{1}{\tau_{w,i}} + \frac{1+K_{p,j}}{\tau_j}}, \tag{4.40}$$

$$S_{3,ij} = \frac{\gamma_i S_{2,ij} + \gamma_j S_{2,ji}}{\frac{1+K_{p,i}}{\tau_i} + \frac{1+K_{p,j}}{\tau_j}}. \tag{4.41}$$

In addition, by utilizing (4.35) and (4.32), J_2 in (4.33) may be stated as

$$J_2 = \text{trace}(S_3 Q_{22}) = \sum_{i=1}^{N}\sum_{j=1}^{N} S_{3,ij} + \sum_{i=1}^{N} \epsilon_i K_{p,i}^2 S_{3,ii}. \tag{4.42}$$

Expressions (4.39)–(4.42) call for a couple of remarks: first, by substituting the $S_{1,ij}$'s from (4.39) into (4.40), the resulting $S_{2,ij}$'s from (4.40) into (4.41), and the resulting $S_{3,ij}$'s from (4.41) into (4.42), we see that the cost function J_2 is an explicit function of the covariance matrix W, time constants $\tau_{w,i}$'s, time constants τ_i's, scalar gains γ_i's, weighting factors ϵ_i's, and proportional controller gains $K_{p,i}$'s, i.e.,

$$\begin{aligned} J_2 = J_2(&W, \tau_{w,1}, \tau_{w,2}, \ldots, \tau_{w,N}, \tau_1, \tau_2, \ldots, \tau_N, \\ &\gamma_1, \gamma_2, \ldots, \gamma_N, \epsilon_1, \epsilon_2, \ldots, \epsilon_N, K_{p,1}, K_{p,2}, \ldots, K_{p,N}). \end{aligned} \tag{4.43}$$

Second, by examining (4.39)–(4.42), we see that the smaller the W_{ij}'s, the smaller the $S_{3,ij}$'s and, thus, J_2. This agrees with the intuition that the weaker the wind speed fluctuations, the smoother the wind farm power output and lower the control effort. We also see that the larger the $K_{p,i}$'s, the smaller the $S_{3,ij}$'s. However, due to the $K_{p,i}^2$'s in (4.42), J_2 might increase or decrease. This again agrees with the intuition that the more aggressive the control actions, the smoother the wind farm power output but a weighted combination of the smoothness and control effort might increase or decrease.

To make use of the above results, recall that the goal of the Optimization of Proportional Controller Gains block is to find $K_{p,i} > 0$ for $i \in \{1, 2, \ldots, N\}$ that minimize J_2 defined in (4.31). To attain this goal, we have converted (4.31) into an explicit form (4.39)–(4.42). We have also listed in (4.43) all the parameters that affect J_2, which can be divided into three groups: (i) the unknown W and $\tau_{w,i}$'s; (ii) the known τ_i's, γ_i's, and ϵ_i's; and (iii) the optimization variables $K_{p,i}$'s. Because of (i)–(iii), by having the block replace the unknown W and $\tau_{w,i}$'s in (4.39)–(4.43) with their presumably accurate estimates $\widehat{W}$ and $\widehat{\tau}_{w,i}$'s, the block would have all the information it needs to compute $K_{p,i}$'s that minimize J_2. As for how the block would go about computing such $K_{p,i}$'s, note from (4.39)–(4.42) that J_2 is a complicated, non-convex function of the $K_{p,i}$'s. Hence, finding a global minimizer may be difficult, especially when N is large. That said, the block could increase its chance of finding a global minimizer by using, for example, a gradient descent optimization algorithm [13] with a diverse set of initial guesses, as follows:

$$\begin{bmatrix} \dot{K}_{p,1} \\ \dot{K}_{p,2} \\ \vdots \\ \dot{K}_{p,N} \end{bmatrix} = -\varepsilon \begin{bmatrix} \frac{\partial J_2}{\partial K_{p,1}} \\ \frac{\partial J_2}{\partial K_{p,2}} \\ \vdots \\ \frac{\partial J_2}{\partial K_{p,N}} \end{bmatrix},$$

where $\varepsilon > 0$ is an algorithm parameter and $\frac{\partial J_2}{\partial K_{p,i}}$ for $i \in \{1, 2, \ldots, N\}$ are partial derivatives of J_2 derivable from (4.39)–(4.42). For this reason, we will assume that such an algorithm is embedded in the Optimization of Proportional Controller Gains block, so that each time it receives periodically available estimates $\widehat{W}$ and $\widehat{\tau}_{w,i}$'s from the Estimation of Wind Speed Parameters block, it could compute near-globally optimal $K_{p,i}$'s and send them to the Proportional Controllers and Feedforward Gains blocks, as depicted in Fig. 4.5. In other words, what it does is it tries to optimally adapt the controller parameters $K_{p,i}$'s to gradual changes in wind speed characteristics as reflected in the estimates $\widehat{W}$ and $\widehat{\tau}_{w,i}$'s. We note that this kind of adaptation is akin to a self-tuning regulator in adaptive control [9], and that is why we call the overall scheme on the inner feedback loop of the WFC an adaptive controller.

4.6 Simulation Results

In this section, we demonstrate the effectiveness of the WFC developed by simulating its model predictive controller in isolation, analyzing its adaptive controller in isolation, and simulating the overall WFC.

4.6.1 Model Predictive Controller in Isolation

In this subsection, we simulate the model predictive controller in isolation. By isolation, we mean that its Reference Models block and the adaptive controller in Fig. 4.5 are assumed to be absent, and the $P^*_{d,i}$'s in the figure are treated as the $P_{d,i}$'s. Our aim here is to show that the model predictive controller is indeed capable of taking advantage of the forecast availability and design freedom mentioned in Characteristics B1 and B2 of Sect. 4.1.

Consider a wind farm with $N = 2$ turbines. For these turbines, let $\alpha_1 = \alpha_2 = 0.657$, $\tau_1 = 1$ min, and $\tau_2 = 1.5$ min. For the wind speeds, let $\overline{V}_{w,1}(t) = 1$ pu, $\overline{V}_{w,2}(t) = 0.8$ pu, and $\widetilde{V}_{w,1}(t) = \widetilde{V}_{w,2}(t) = 0$. For the model predictive controller, let $T = 200$ min, $T_s = 1$ min, $\eta(k) = 1$, $\mu_1(k) = 1$, $\mu_2(k) = 2$, $\nu_1(k) = \nu_2(k) = 10^{-6}$, $\widehat{V}_{w,1}(t) = \overline{V}_{w,1}(t)$, $\widehat{V}_{w,2}(t) = \overline{V}_{w,2}(t)$, $P_1[0] = 0.2$ pu, $P_2[0] = 0.8$ pu, $P_{d,1}[0] = 0.55$ pu, and $P_{d,2}[0] = 0.35$ pu. Note that since the adaptive controller is not present to reject the $\widetilde{V}_{w,i}(t)$'s, we let $\widetilde{V}_{w,1}(t) = \widetilde{V}_{w,2}(t) = 0$ here, and will only study their rejection when the adaptive controller is present, in the next two subsections. In addition, since $\widetilde{V}_{w,1}(t) = \widetilde{V}_{w,2}(t) = 0$, γ_1, γ_2, W, $\tau_{w,1}$, and $\tau_{w,2}$ need not be specified.

Next, consider two scenarios. For the first scenario, let $P_{d,wf}(t) = 0.5$ pu for $t \in [0, 30)$ and $P_{d,wf}(t) = 1$ pu for $t \in [30, \infty)$, where t is in minutes. Moreover, let the forecast of $P_{d,wf}(t)$ be accurate. For this scenario, subplot (1, 1) of Fig. 4.6 shows the simulation result produced by *one-shot optimization*, i.e., by solving optimization problem (4.12) once, at time $t = 0$, and applying the entire solution. In contrast, subplot (2, 1) shows the simulation result produced by *iterative optimization* within the model predictive controller, i.e., by solving (4.12) iteratively and applying only the first step of the solution. Observe that these two subplots have little differences, which is not surprising since the forecasts of $P_{d,wf}(t)$, $\overline{V}_{w,1}(t)$, and $\overline{V}_{w,2}(t)$ are all accurate. Also notice that, due to utilization of these forecasts, the model predictive controller is able to start taking actions before changes actually occur. That is, it is able to anticipate a surge in $P_{d,wf}(t)$ from 0.5 pu to 1 pu at time $t = 30$, and start increasing $P_{d,1}(t)$ and $P_{d,2}(t)$ beforehand at time $t \approx 26$, so that $P_{wf}(t)$ more accurately and gracefully tracks $P_{d,wf}(t)$. We note that this look-ahead ability is not possessed by existing WFCs in [32, 40, 50, 70, 81, 93, 101, 103, 104].

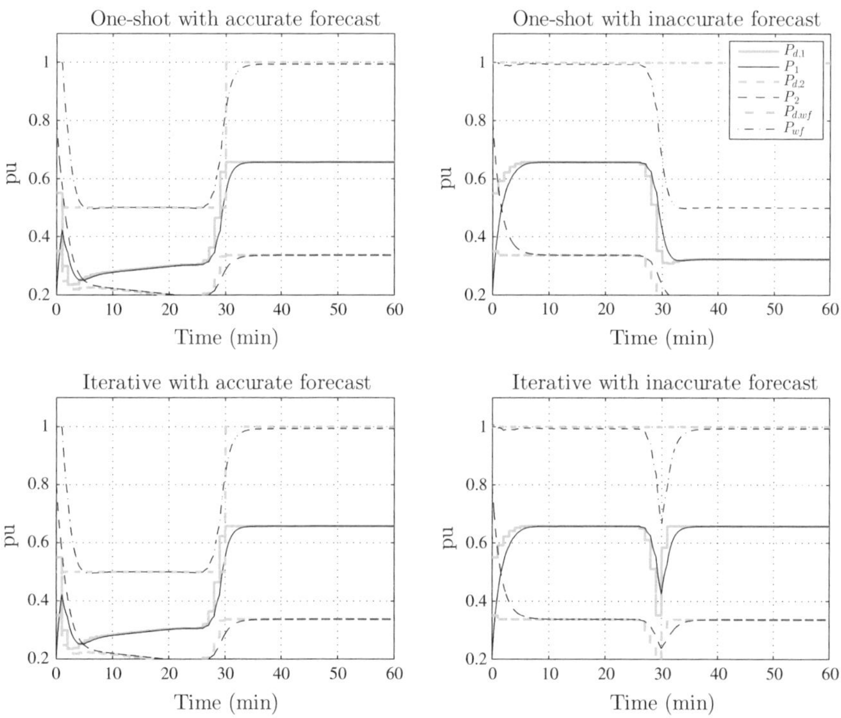

Fig. 4.6 Model predictive controller in isolation: One-shot optimization versus iterative optimization

For the second scenario, let $P_{d,wf}(t) = 1$ pu for $t \in [0, \infty)$. Furthermore, for any t_1 and t_2 with $0 \leq t_1 \leq t_2 < \infty$, let $\widehat{P}_{d,wf}(t_1, t_2)$ denote the forecast made at time t_1 of the value of $P_{d,wf}(t_2)$ at time t_2, defined as

$$\widehat{P}_{d,wf}(t_1, t_2) = \begin{cases} 0.5, & \text{if } t_1 \in [0, 30) \text{ and } t_2 \in [30, \infty), \\ 1, & \text{otherwise}. \end{cases} \tag{4.44}$$

Equation (4.44) says that: (i) the forecast made at any time within the first 30 min points to a drop in $P_{d,wf}(t)$ from 1 pu to 0.5 pu at time $t = 30$, which turns out to be inaccurate as such a drop never materializes; and (ii) the forecast made at any time after the first 30 min points to a constant $P_{d,wf}(t)$ of 1 pu, which turns out to be accurate. For this scenario, subplot (1, 2) of Fig. 4.6 shows the simulation result produced by *one-shot optimization*, while subplot (2, 2) shows the one produced by *iterative optimization* within the model predictive controller. Observe from time interval [25, 30] of these two subplots that both the one-shot and iterative optimization schemes are misled by the inaccurate forecast in (i). However, as is evident from time interval [30, 40], the latter is able to quickly react to the revised, accurate forecast in (ii), whereas the former is not. We remark that this ability of the model predictive controller is also not possessed by the existing WFCs.

4.6.2 Adaptive Controller in Isolation

In this subsection, we analyze the adaptive controller in isolation. By isolation, we mean that the model predictive controller in Fig. 4.5 is assumed to be absent, and the P_i^*'s in the figure are treated as given constants. Our aim here is to show that the adaptive controller is indeed capable of exploiting the wind speed correlation mentioned in Characteristic B3 of Sect. 4.1.

Consider a wind farm with $N = 4$ turbines. For these turbines, let $\tau_i = 1$ min and $\gamma_i = 1$ for all i. For the wind speeds, let $\widetilde{V}_{w,i}(t)$ be described by (4.19), W be as defined below, and $\tau_{w,i} = 5$ s for all i. For the adaptive controller, let $\epsilon_i = \epsilon$, $\widehat{W} = W$, $\widehat{\tau}_{w,i} = \tau_{w,i}$, and $K_{p,i}$ be computed according to Sect. 4.5.3.3 for all i. Notice from the development in Sect. 4.5.3 that the α_i's, $\overline{V}_{w,i}(t)$'s, and P_i^*'s do not matter and, therefore, need not be specified.

Next, consider the following four scenarios that yield four different wind speed correlations W's:

- *Moderately correlated*: For the first scenario, portrayed in Fig. 4.7a, the turbines are lined up horizontally from west to east, and face a wind that blows from the west. In addition, the turbines are placed in a way that the distances between them are small, but still large enough that their wake effects do not affect one another. With such a turbine placement and such a wind, the correlation between $\widetilde{V}_{w,i}(t)$ and $\widetilde{V}_{w,j}(t)$ is likely to be high if turbines i and j are adjacent, and low if they are far apart. Thus, a plausible W for this "moderately correlated" scenario is

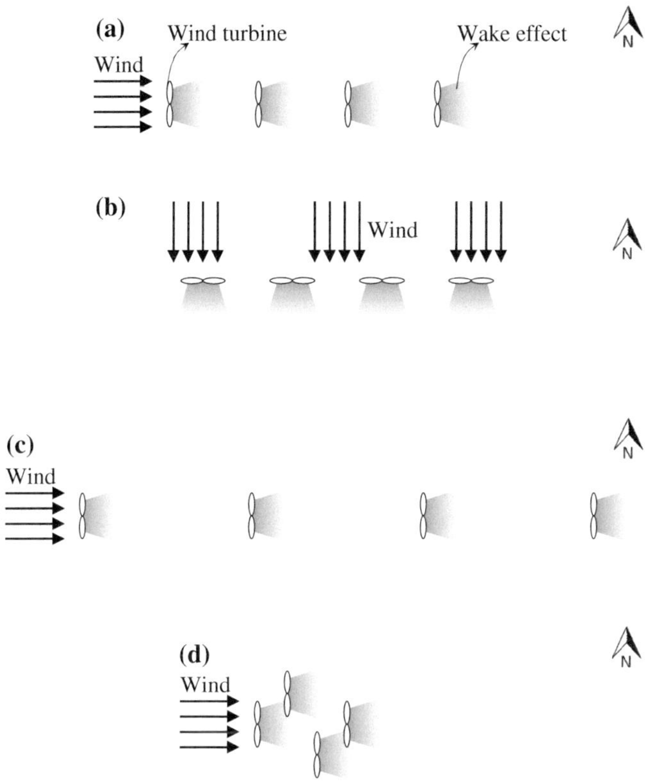

Fig. 4.7 Adaptive controller in isolation: Four scenarios that yield four different wind speed correlations W's. **a** Moderately correlated. **b** Strongly correlated. **c** Totally uncorrelated. **d** Negatively correlated

$$W = \begin{bmatrix} 1.0 & 0.85 & 0.5 & 0.0 \\ 0.85 & 1.0 & 0.85 & 0.5 \\ 0.5 & 0.85 & 1.0 & 0.85 \\ 0.0 & 0.5 & 0.85 & 1.0 \end{bmatrix}.$$

- *Strongly correlated*: For the second scenario, depicted in Fig. 4.7b, the turbines are also lined up horizontally from west to east, but face a wind that blows from the north. Assuming that the wind from the north is created by pressure gradients that are nearly constant horizontally, the correlation between $\widetilde{V}_{w,i}(t)$ and $\widetilde{V}_{w,j}(t)$ is likely to be high if turbines i and j are adjacent, and somewhat high if they are far apart. Hence, a reasonable W for this "strongly correlated" scenario is

$$W = \begin{bmatrix} 1.0 & 0.9 & 0.8 & 0.7 \\ 0.9 & 1.0 & 0.9 & 0.8 \\ 0.8 & 0.9 & 1.0 & 0.9 \\ 0.7 & 0.8 & 0.9 & 1.0 \end{bmatrix}.$$

- *Totally uncorrelated*: The third scenario, portrayed in Fig. 4.7c, is identical to the first scenario in Fig. 4.7a, except that the distances between the turbines are very large. Due to that, the correlation between $\widetilde{V}_{w,i}(t)$ and $\widetilde{V}_{w,j}(t)$, for any $i \neq j$, is likely to be negligible. Therefore, a plausible W for this "totally uncorrelated" scenario is

$$W = \begin{bmatrix} 1.0 & 0.0 & 0.0 & 0.0 \\ 0.0 & 1.0 & 0.0 & 0.0 \\ 0.0 & 0.0 & 1.0 & 0.0 \\ 0.0 & 0.0 & 0.0 & 1.0 \end{bmatrix}.$$

- *Negatively correlated*: In the above three scenarios, the wind speed correlation is either strong, moderate, or nonexistent. Moreover, the entries of W are either positive or zero, even though they are allowed to be negative (as long as W remains symmetric positive definite). Thus, to cover more ground we let the fourth scenario be a "negatively correlated" scenario, in which certain entries of W are negative. Specifically, we let

$$W = \begin{bmatrix} 1.0 & -0.3 & -0.3 & -0.3 \\ -0.3 & 1.0 & -0.3 & -0.3 \\ -0.3 & -0.3 & 1.0 & -0.3 \\ -0.3 & -0.3 & -0.3 & 1.0 \end{bmatrix}$$

  and note that this W may be the result of an overcrowded wind farm, where the $\widetilde{V}_{w,i}(t)$'s experienced by the turbines are severely affected by the wake effects from neighboring turbines, as illustrated in Fig. 4.7d.

To analyze the adaptive controller's performance in the above four scenarios, note that because $\epsilon_i = \epsilon$ for all i, J_2 in (4.42) may be viewed as a weighted sum of a *smoothness* term and a *control effort* term, i.e.,

$$J_2 = \underbrace{\sum_{i=1}^{4}\sum_{j=1}^{4} S_{3,ij}}_{\text{Smoothness}} + \epsilon \underbrace{\sum_{i=1}^{4} K_{p,i}^2 S_{3,ii}}_{\text{Control effort}}. \tag{4.45}$$

Furthermore, because the $S_{3,ij}$'s and $S_{3,ii}$'s in (4.45) are functions of W and the $K_{p,i}$'s, J_2 in (4.45) is a function of W, ϵ, and the $K_{p,i}$'s. Hence, for any given W and ϵ, we could compute $K_{p,i}$'s that minimize J_2 and use these $K_{p,i}$'s to compute the smoothness and control effort terms in (4.45). These smoothness and control effort terms would then represent the adaptive controller's performance for the given

W and ϵ. It follows that by performing such a computation for each W from the four scenarios and each ϵ from the interval $(0, \infty)$, we could evaluate the adaptive controller's performance in the four scenarios over a broad range of weighting factors.

Figure 4.8 shows the result of such an evaluation. Specifically, Fig. 4.8a considers the moderately correlated scenario and shows that, as ϵ goes from 0 to ∞ (i.e., goes from cheap control to expensive control), the smoothness term increases from below 50 to 280, while the control effort term decreases from above 400 to 0, tracing out a curve that travels from right to left. Figures 4.8b–d consider the strongly correlated, totally uncorrelated, and negatively correlated scenarios and display similar curves. In each of these figures, the curve is labeled *Pareto optimal front* because it is precisely the Pareto optimal curve of a multi-objective optimization problem whose two objective functions are the smoothness and control effort terms in (4.45), and whose optimization variables are the $K_{p,i}$'s. For this reason, the curve is shown to separate the plane into an achievable region which is marked in gray, and an unachievable one which is not attainable by all $K_{p,i}$'s.

Analyzing Fig. 4.8, we can make the following observations: first, the adaptive controller is always Pareto optimal. This is because it is able to operate at a point on one of the four Pareto optimal curves for every scenario (i.e., every wind speed correlation W) and every weighting factor ϵ. Notice that if its $K_{p,i}$'s were not chosen

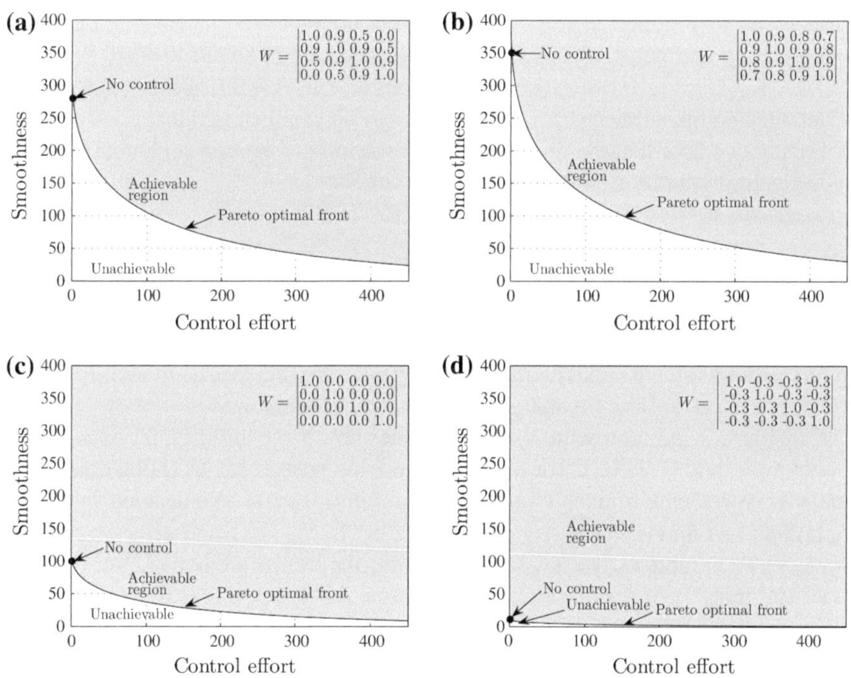

Fig. 4.8 Adaptive controller in isolation: Pareto optimal curves for the four scenarios. **a** Moderately correlated. **b** Strongly correlated. **c** Totally uncorrelated. **d** Negatively correlated

to minimize J_2, it would have been suboptimal due to operating in the interior of one of the four achievable regions.

Second, there is a trade-off between smoothness and control effort in every scenario. In particular, if the adaptive controller is unwilling to put in any control effort, it would operate at the point labeled *No control* and experience the worst possible smoothness. On the contrary, if it is willing to put in arbitrarily large control effort, it could enjoy arbitrarily good smoothness.

Third, the adaptive controller is capable of coping with changes in the wind speed correlation W. For instance, if the wind direction gradually changes from eastward in the moderately correlated scenario of Fig. 4.7a, to southward in the strongly correlated scenario of Fig. 4.7b, the controller is able to adapt its $K_{p,i}$'s to changes in W to remain Pareto optimal.

Fourth, the adaptive controller performs best in the negatively correlated scenario, followed by the totally uncorrelated, moderately correlated, and strongly correlated scenarios. This ranking of the scenarios is based on their levels of smoothness at a given level of control effort (which may be any level). The ranking makes intuitive sense because the strongly correlated scenario tends to produce $w_1(t), w_2(t), w_3(t), w_4(t)$ that are "one-sided," i.e., either all positive or all negative. These one-sided $w_i(t)$'s tend to yield one-sided $\widetilde{V}_{w,1}(t), \widetilde{V}_{w,2}(t), \widetilde{V}_{w,3}(t), \widetilde{V}_{w,4}(t)$ via (4.19), and one-sided $\Delta P_1(t), \Delta P_2(t), \Delta P_3(t), \Delta P_4(t)$ via (4.27), causing $\sum_{i=1}^{4} \Delta P_i(t)$ to have a large variability. In contrast, the negatively correlated scenario tends to produce $w_1(t), w_2(t), w_3(t), w_4(t)$ that are "balanced," i.e., have both positive and negative values. These balanced $w_i(t)$'s tend to yield balanced $\widetilde{V}_{w,1}(t), \widetilde{V}_{w,2}(t), \widetilde{V}_{w,3}(t), \widetilde{V}_{w,4}(t)$ and balanced $\Delta P_1(t), \Delta P_2(t), \Delta P_3(t), \Delta P_4(t)$ that offset one another, allowing $\sum_{i=1}^{4} \Delta P_i(t)$ to have a small variability.

Finally, we note that the result in this subsection is obtained analytically (albeit by neglecting saturation) instead of through simulation.

4.6.3 Overall Wind Farm Controller

In this subsection, we simulate the overall WFC, showing that its model predictive and adaptive controllers are able to work together effectively.

Consider a wind farm with $N = 10$ turbines. For these turbines, let $\alpha_i = 0.657$, $\tau_i = 1$ min, and $\gamma_i = 0.02$ for all i. For the wind speeds, let $\mathsf{V_w}(t)$ be defined by real wind speed data from an Oklahoma wind farm, $\overline{V}_{w,i}(t)$ be a delayed version of $\mathsf{V_w}(t)$ by $i - 1$ min (i.e., $\overline{V}_{w,i}(t) = \mathsf{V_w}(t - i + 1)$), $\widetilde{V}_{w,i}(t)$ be described by (4.19), $W_{ij} = 0.9^{|i-j|}$, and $\tau_{w,i} = 1$ s for all i, j. For the model predictive controller, let $T = 100$ min, $T_s = 1$ min, $\eta(k) = 1$, $\nu_i(k) = 10^{-6}$, $\widehat{V}_{w,i}(t) = \overline{V}_{w,i}(t)$, $P_i[0] = 0.2$ pu, and $P_{d,i}[0] = 0.2$ pu for all i. In addition, let $\mu_1(k) = \mu_6(k) = 1$, $\mu_2(k) = \mu_7(k) = 2$, $\mu_3(k) = \mu_8(k) = 4$, $\mu_4(k) = \mu_9(k) = 8$, and $\mu_5(k) = \mu_{10}(k) = 16$, so that changes in the $P_{d,i}[k]$'s are least penalized for turbines 1 and 6, and most penalized for turbines 5 and 10. For the adaptive controller, let $\epsilon_i = 1$, $\widehat{W} = W$, $\widehat{\tau}_{w,i} = \tau_{w,i}$, and $K_{p,i}$ be computed according to Sect. 4.5.3.3 for all i. As for the power grid operator, let $P_{d,wf}(t) = 2$ pu for $t \in [0, 20)$, $P_{d,wf}(t) = 4$ pu for $t \in [20, 40)$, and $P_{d,wf}(t) = 2$ pu for $t \in [40, \infty)$, so that $P_{d,wf}(t)$ experiences jumps at times $t = 20$ and $t = 40$. Moreover, let the forecast of $P_{d,wf}(t)$ be accurate.

Figures 4.9, 4.10 and 4.11 display the simulation result. Specifically, Fig. 4.9 shows the wind speeds $V_{w,i}(t)$'s as functions of time t, Fig. 4.10 shows the wind turbines' desired power references $P_{d,i}(t)$'s and actual power outputs $P_i(t)$'s also as functions of time t, and Fig. 4.11 does the same for the wind farm's desired power output $P_{d,wf}(t)$ and actual power output $P_{wf}(t)$. Observe from Fig. 4.10 that $P_{d,1}(t)$ and $P_{d,6}(t)$ experience the most changes, whereas $P_{d,5}(t)$ and $P_{d,10}(t)$ experience the least. Furthermore, every $P_i(t)$ is able to closely track $P_{d,i}(t)$, despite the fast fluctuations in $V_{w,i}(t)$ from Fig. 4.9. The first observation may be attributed to changes in the $P_{d,i}(t)$'s being penalized differently, while the second observation may be attributed to the turbine inertias and to the proportional controllers being well-designed. Lastly, notice from Fig. 4.11 that $P_{wf}(t)$ is able to accurately and smoothly track $P_{d,wf}(t)$, achieving the ultimate objective of the WFC. This also implies that its model predictive and adaptive controllers are able to work together effectively, with the former ensuring accurate tracking on a longer timescale, and the latter ensuring smoothness on a shorter one.

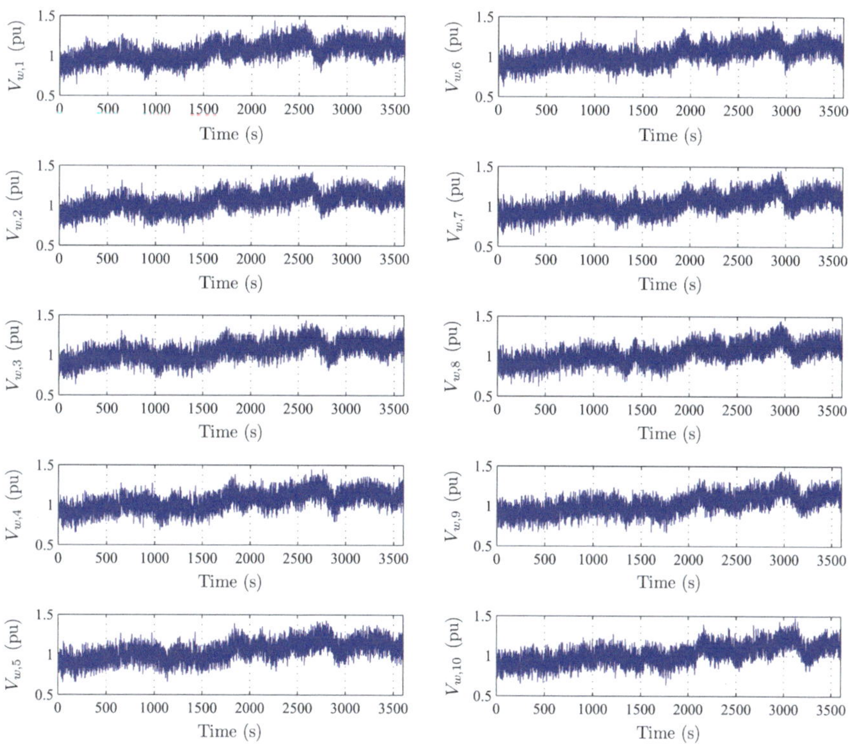

Fig. 4.9 Overall WFC: Wind speeds $V_{w,i}(t)$'s used in the simulation

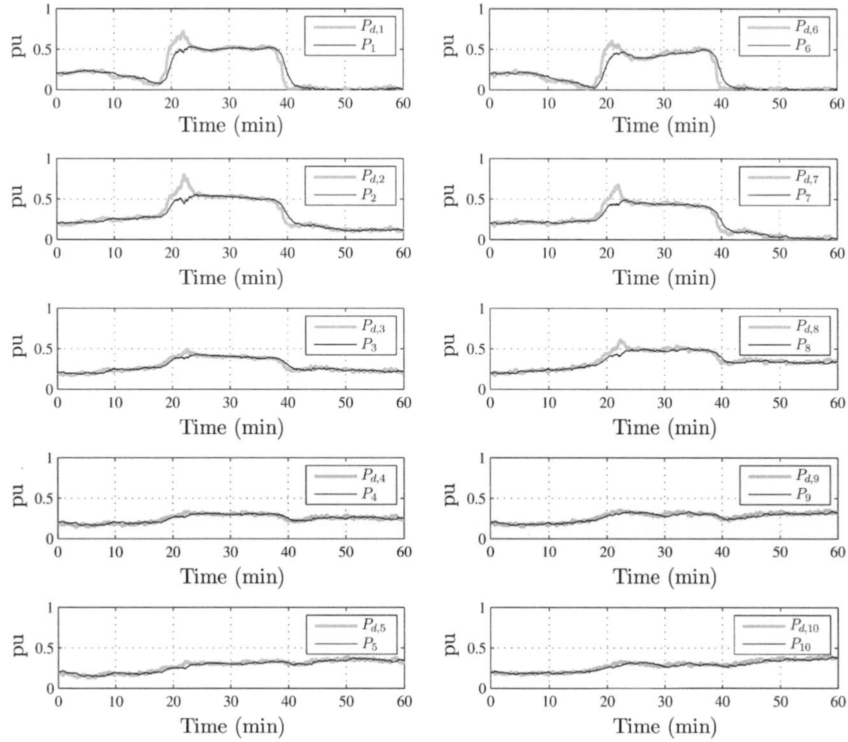

Fig. 4.10 Overall WFC: Wind turbines' desired power references $P_{d,i}(t)$'s and actual power outputs $P_i(t)$'s generated by the simulation

Fig. 4.11 Overall WFC: Wind farm's desired power output $P_{d,wf}(t)$ and actual power output $P_{wf}(t)$ generated by the simulation

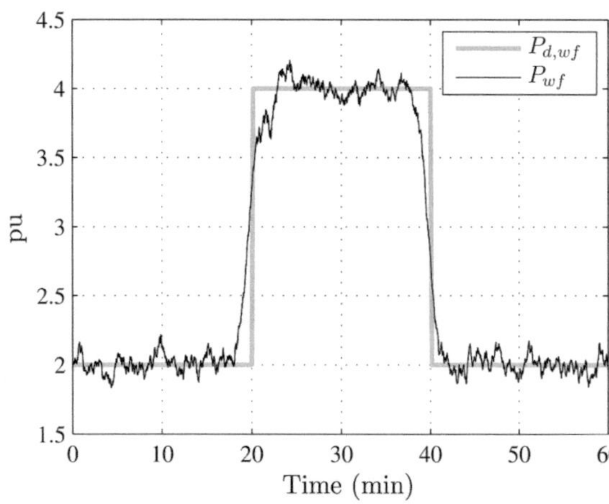

4.7 Concluding Remarks

In this chapter, we have developed a novel supervisory WFC that enables the active power output of a wind farm to accurately and smoothly track a desired reference provided by a power grid operator. In the process of developing this WFC, we have explained the rationale behind its two-loop architecture. We have also shown that the outer feedback loop of the WFC contains a model predictive controller, which uses various forecasts and feedbacks to iteratively compute a set of desired power trajectories, so that the deterministic tracking accuracy of the wind farm power output on a receding horizon is optimized. In contrast, the inner feedback loop contains an adaptive controller, which uses estimated wind speed characteristics to adaptively tune a set of proportional controller gains, so that the stochastic smoothness of the wind farm power output on a shorter timescale is optimized. Lastly, we have carried out a series of simulation studies that demonstrate the effectiveness of the WFC, including its ability to exploit forecast availability, design freedom, and wind speed correlation.

Chapter 5
Quasilinear Control of Wind Farm Active Power Output

Abstract *Quasilinear Control* (QLC) is a set of methods for analysis and design of systems with nonlinear actuators and sensors. The approach of QLC is based on the method of stochastic linearization, which replaces each nonlinearity with an equivalent gain and bias. In this chapter, we revisit the wind farm controller from Chap. 4 and apply QLC to the design of the proportional controller gains on its inner feedback loop. QLC is particularly suitable here because each of the wind turbines can be modeled as a linear plant preceded by an asymmetric saturation nonlinearity, which accounts for the limited availability of wind. Through extensive numerical simulation, we show that controller gains designed using QLC perform significantly better in a broad range of wind farm operating regimes, compared to those designed using traditional linear methods, where the asymmetric saturation nonlinearities are ignored. The operating regimes we consider include regimes with low and high wind speed, medium and high power generation, weak and strong wind correlation, and cheap and expensive control.

5.1 Introduction

In Chap. 3, we present a simple approximate model that is capable of closely mimicking the active and reactive power dynamics of various *wind turbine control systems* (WTCSs) under normal operating conditions. Based on this WTCS model, we develop in Chap. 4 a *wind farm controller* (WFC) consisting of a model predictive controller and an adaptive controller, which collectively enable the wind farm power output to accurately and smoothly track a desired reference from a power grid operator. Although this WFC has several positive features, it unfortunately also has a notable drawback: in order to use linear control techniques to design the proportional controller gains in the adaptive controller, we assume that the saturation function in the WTCS model is not present, so that the model is completely linear. While this assumption greatly simplifies the design process and subsequent analysis, the results obtained may be overly optimistic and may not accurately reflect the WFC performance when the saturation function is present.

J.N. Jiang et al., *Control and Operation of Grid-Connected Wind Farms*,
Advances in Industrial Control, DOI 10.1007/978-3-319-39135-9_5

To alleviate this drawback, in this chapter we leverage the recently-developed theory of *Quasilinear Control* (QLC) [29, 85]. QLC is a set of methods for design and analysis of control systems with static nonlinearities in their actuators and sensors. Such systems are referred to as *linear plant/nonlinear instrumentation* (LPNI) systems. The methods in QLC are based on stochastic linearization [92], which replaces each nonlinearity in an LPNI system with an equivalent gain and an equivalent bias, to arrive at a quasilinear system. Unlike the usual Jacobian linearization which is inherently local, stochastic linearization is global. Moreover, as shown in [29, 85], the resulting quasilinear system is able to closely approximate signals in the original LPNI system if the latter has sufficient low-pass filtering characteristics—which, as we will show, is the case with the WTCS model considered here. Furthermore, since the QLC method takes into account the saturation nonlinearity in the design stage, it is expected to yield a better performance compared to the linear method in Chap. 4 which neglects saturation.

In Sect. 5.2 of this chapter, we review the *wind farm control system* (WFCS) from Chap. 4, including only the details that are necessary later. We also state the problem to be addressed. In Sect. 5.3, we use the QLC method to design the proportional controller gains in the adaptive controller. Upon completion, we demonstrate via extensive numerical simulation in Sect. 5.4 that the QLC method indeed significantly outperforms the linear method in Chap. 4. Specifically, the QLC method is no worse, and up to 60 % better, than the linear method in all the operating regimes considered, including regimes with low and high wind speed, medium and high power generation, weak and strong wind correlation, and cheap and expensive control. Hence, the main contribution of the chapter is the novel and successful application of QLC theory to wind farm active power control, which facilitates the integration of wind farms into power grids.

5.2 Model and Problem Formulation

5.2.1 *Wind Farm Control System*

Consider a WFCS consisting of N WTCSs and a WFC, as depicted in Fig. 4.1. Suppose each WTCS $i \in \{1, 2, \ldots, N\}$ is made up of a wind turbine and its turbine-level controller, which we assume have both been specified. Also suppose each WTCS i accepts as inputs a wind speed $V_{w,i}(t) \in \mathbb{R}$ and a desired power reference $P_{d,i}(t) \in \mathbb{R}$ and produces as output an actual power $P_i(t) \in \mathbb{R}$, where $t \geq 0$ denotes time. Moreover, suppose the following information is available to the WFC at each time t: measurements of the $P_i(t)$'s, estimates of the $V_{w,i}(t)$'s, a desired wind farm power output $P_{d,wf}(t) \in \mathbb{R}$, and a minutes-to-hours-ahead forecast of $P_{d,wf}(t)$, the latter two from a power grid operator. Given the N WTCSs and such information, the goal of the WFC is to calculate the control inputs $P_{d,i}(t)$'s, so that the actual wind farm power output $P_{wf}(t) \triangleq \sum_{i=1}^{N} P_i(t)$ accurately and smoothly tracks $P_{d,wf}(t)$, despite the fluctuating $V_{w,i}(t)$'s. To facilitate the development of such a WFC, we describe in Sects. 5.2.2–5.2.4 a model for the N WTCSs, followed by an existing WFC design and a precise problem statement.

5.2.2 Wind Turbine Control System Model

As explained in Chap. 3, the design and analysis of a WFC can be challenging because the dynamics of a large number of WTCSs are complex. To reduce complexity, a WTCS model is developed in the chapter based on standard system identification approaches and typical WTCS characteristics. Composed of a static nonlinear element and a linear time-invariant system, this structurally simple model is shown to be accurate and versatile. Indeed, by letting its parameters take different values, the model is capable of closely imitating the active and reactive power dynamics of several different analytical and empirical WTCS models from the literature and from real data, including those in [40, 59, 93, 106]. In the current chapter, we utilize this WTCS model in conjunction with a simple wind speed model in essentially the same way as is done in Chap. 4. The two models are described below.

With the wind speed model, the wind speed $V_{w,i}(t)$ entering each WTCS $i \in \{1, 2, \ldots, N\}$ is given by

$$V_{w,i}(t) = \overline{V}_{w,i}(t) + \widetilde{V}_{w,i}(t), \tag{5.1}$$

where $\overline{V}_{w,i}(t) > 0$ represents the slow, average component of $V_{w,i}(t)$, and $\widetilde{V}_{w,i}(t) \in \mathbb{R}$ represents the fast, deviation-from-average component of $V_{w,i}(t)$. Each slow component $\overline{V}_{w,i}(t)$ is assumed to be deterministic and specified by either empirical data (e.g., historical hourly weather data) or test signals (e.g., step, ramp, sinusoidal). In contrast, each fast component $\widetilde{V}_{w,i}(t)$ is assumed to be a stationary, zero-mean colored Gaussian random process specified by

$$\dot{\widetilde{V}}_{w,i}(t) = -\frac{1}{\tau_{w,i}}\widetilde{V}_{w,i}(t) + \frac{1}{\tau_{w,i}}w_i(t), \tag{5.2}$$

where $\tau_{w,i} > 0$ is the time constant, $w(t) = (w_1(t), \ldots, w_N(t)) \in \mathbb{R}^N$ is a stationary, zero-mean white Gaussian random process with autocovariance function $\mathrm{E}\{w(t)w^T(\tau)\} = W\delta(t-\tau)$, E is the expectation operator, $W = W^T > 0$ is the covariance matrix, and δ is the Dirac delta function. Notice that since $\widetilde{V}_{w,i}(t)$ is Gaussian, despite $\overline{V}_{w,i}(t)$ being positive, $V_{w,i}(t)$ in (5.1) may be negative with a small probability. For simplicity, however, we will allow that in this chapter. Also note that in reality, wind speed is a nonstationary random process due to changes in wind directions, weather conditions, and turbine yaw angles. Such changes, however, are usually very slow, so that $\widetilde{V}_{w,i}(t)$ in (5.2) may be considered stationary.

With the WTCS model, the dynamics of each WTCS $i \in \{1, 2, \ldots, N\}$ are given by

$$\dot{P}_i(t) = -\frac{1}{\tau_i} P_i(t) + \frac{1}{\tau_i} \operatorname{sat}_0^{\alpha_i \overline{V}_{w,i}^3(t)}(P_{d,i}(t)) + \gamma_i \widetilde{V}_{w,i}(t), \tag{5.3}$$

where $\tau_i > 0$ is the time constant, $\alpha_i > 0$ is a unit conversion factor, $\gamma_i \geq 0$ is a scalar gain, and $\operatorname{sat}_a^b(u) \triangleq \max\{\min\{u, b\}, a\}$ is a unit-slope, asymmetric saturation function with limits a and b, as illustrated in Fig. 5.1. To understand the dynamics (5.3), first recall that each WTCS i is made up of a wind turbine and its turbine-level controller. Next suppose the wind speed $V_{w,i}(t)$ and desired power reference $P_{d,i}(t)$ are both constant, i.e., $V_{w,i}(t) = V_{w,i}$ and $P_{d,i}(t) = P_{d,i}$, so that $\overline{V}_{w,i}(t) = V_{w,i}$ and $\widetilde{V}_{w,i}(t) = 0$. Then, the maximum power the turbine can produce is a constant given by $\alpha_i V_{w,i}^3$, where $\alpha_i = \frac{1}{2}\rho A_i C_{p,i}^{\max}$, ρ is the air density, A_i is its swept area, and $C_{p,i}^{\max}$ is the maximum value of its performance coefficient. Thus:

- If $P_{d,i} > \alpha_i V_{w,i}^3$ (i.e., the desired power exceeds what the turbine can produce) and if the turbine-level controller is well-designed (to operate in the so-called *maximum power tracking* mode), the actual power output $P_i(t)$ would gradually approach $\alpha_i V_{w,i}^3$. This behavior is captured by (5.3): if $P_{d,i} > \alpha_i V_{w,i}^3$, (5.3) reduces to

$$\dot{P}_i(t) = -\frac{1}{\tau_i} P_i(t) + \frac{1}{\tau_i} \alpha_i V_{w,i}^3,$$

  so that $P_i(t)$ asymptotically converges to $\alpha_i V_{w,i}^3$ at a rate depending on τ_i.
- If, instead, $0 \leq P_{d,i} \leq \alpha_i V_{w,i}^3$ (i.e., the desired power is producible) and if the controller is well-designed (to operate in the *power regulation* mode), $P_i(t)$ would go to $P_{d,i}$. This behavior is also captured by (5.3) as it becomes

$$\dot{P}_i(t) = -\frac{1}{\tau_i} P_i(t) + \frac{1}{\tau_i} P_{d,i},$$

  forcing $P_i(t)$ to go to $P_{d,i}$.

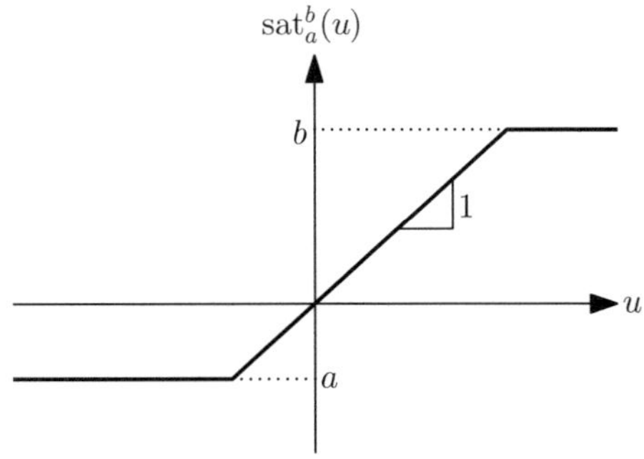

Fig. 5.1 Asymmetric saturation function

- If, instead, $P_{d,i} < 0$ (which means the WFC somehow wants the turbine to act as a motor) and if the controller is well-designed (to prevent that), $P_i(t)$ might decrease but would remain positive. This behavior is again captured by (5.3) as it becomes

$$\dot{P}_i(t) = -\frac{1}{\tau_i} P_i(t).$$

- Finally, if $V_{w,i}(t)$ and $P_{d,i}(t)$ are time-varying, $\alpha_i \overline{V}_{w,i}^3(t)$ would represent the maximum producible power, $\widetilde{V}_{w,i}(t)$ would create fluctuations in $P_i(t)$, and $P_i(t)$ would try to follow $P_{d,i}(t)$. These three behaviors are modeled, respectively, by the upper saturation limit $\alpha_i \overline{V}_{w,i}^3(t)$, the disturbance term $\gamma_i \widetilde{V}_{w,i}(t)$, and the first-order dynamics with time constant τ_i, in (5.3).

Therefore, although simple, model (5.3) is able to capture several essential characteristics of a well-designed WTCS.

5.2.3 *Existing Wind Farm Controller Design*

Based on (5.1)–(5.3), a WFC is designed in Chap. 4, which strives to make the wind farm power output $P_{wf}(t)$ track the desired $P_{d,wf}(t)$ by adjusting the control inputs $P_{d,i}(t)$'s. In this subsection, we briefly summarize the WFC, providing only the details that are needed in subsequent sections.

The WFC consists of an outer loop and an inner loop. The outer loop comprises a *model predictive controller*, whose goal is to optimize the deterministic tracking accuracy of $P_{wf}(t)$ on a receding horizon. To achieve its goal, the model predictive controller uses estimates of the $V_{w,i}(t)$'s, a forecast of $P_{d,wf}(t)$, and measurements of the $P_i(t)$'s to iteratively determine a set of N intermediate power references $P_i^*(t)$'s, which drive the inner loop. The inner loop comprises an *adaptive controller*, whose goal is to optimize the steady-state, stochastic smoothness of $P_{wf}(t)$ on a short timescale. To accomplish its goal, the adaptive controller uses measurements of the $V_{w,i}(t)$'s to continuously estimate the covariance matrix W of $w(t)$ and time constants $\tau_{w,i}$'s in (5.2), which are then used to continuously optimize the positive gains $K_{p,i}$'s of N decoupled proportional controllers. The proportional controllers, in turn, use knowledge of the $P_i^*(t)$'s and $K_{p,i}$'s and measurements of the $P_i(t)$'s to calculate the control inputs $P_{d,i}(t)$'s, which drive the N WTCSs. Specifically, the control law for each WTCS $i \in \{1, 2, \dots, N\}$ is given by

$$P_{d,i}(t) = K_{p,i}\left(\frac{1 + K_{p,i}}{K_{p,i}} P_i^*(t) - P_i(t)\right), \tag{5.4}$$

where $\frac{1+K_{p,i}}{K_{p,i}}$ is a feedforward gain intended to yield an appropriate equilibrium point.

Substituting control law (5.4) into WTCS model (5.3) and assuming that the slow wind speed components $\overline{V}_{w,i}(t)$'s and intermediate power references $P_i^*(t)$'s are so slow that they may be treated as constants, we obtain for each $i \in \{1, 2, \ldots, N\}$

$$\dot{P}_i(t) = -\frac{1}{\tau_i} P_i(t) + \frac{1}{\tau_i} \operatorname{sat}_0^{\alpha_i \overline{V}_{w,i}^3} \left(K_{p,i} \left(\frac{1 + K_{p,i}}{K_{p,i}} P_i^* - P_i(t) \right) \right) + \gamma_i \widetilde{V}_{w,i}(t). \tag{5.5}$$

Notice that when saturation is not activated and when there is no "disturbance" $\widetilde{V}_{w,i}(t)$, system (5.5) has a unique equilibrium point at $P_i(t) = P_i^*$, which is desirable. Also note from (5.4) that at this equilibrium point, the corresponding control input is $P_{d,i}(t) = P_i^*$. To translate both the equilibrium point and the corresponding control input to the origin, let us introduce for each $i \in \{1, 2, \ldots, N\}$ a new state variable $\Delta P_i(t) \in \mathbb{R}$ and a new control input $\Delta P_{d,i}(t) \in \mathbb{R}$, defined as

$$\begin{aligned} \Delta P_i(t) &= P_i(t) - P_i^*, \\ \Delta P_{d,i}(t) &= P_{d,i}(t) - P_i^*. \end{aligned} \tag{5.6}$$

With (5.6), control law (5.4) becomes

$$\Delta P_{d,i}(t) = -K_{p,i} \Delta P_i(t), \tag{5.7}$$

and system (5.5) can be written as

$$\Delta \dot{P}_i(t) = -\frac{1}{\tau_i} \Delta P_i(t) + \frac{1}{\tau_i} \operatorname{sat}_{-P_i^*}^{\alpha_i \overline{V}_{w,i}^3 - P_i^*} (\Delta P_{d,i}(t)) + \gamma_i \widetilde{V}_{w,i}(t), \tag{5.8}$$

where we have used the fact that $\operatorname{sat}_a^b(u) = a + \operatorname{sat}_0^{b-a}(u - a)$.

Observe that system (5.8) is subject to N decoupled asymmetric saturation functions, whose limits may be different from one another and can change slowly over time. Given that saturation is generally difficult to handle—not to mention multiple asymmetric ones—to facilitate the design and analysis of the said WFC, it is assumed in Chap. 4 that such saturation is not present, with the understanding that the results obtained may be more optimistic than reality. With this assumption, (5.2), (5.7), and (5.8) can be combined to form a linear system driven by white Gaussian noise, i.e.,

$$\begin{bmatrix} \dot{\widetilde{V}}_w(t) \\ \Delta \dot{P}(t) \end{bmatrix} = \underbrace{\begin{bmatrix} A_{11} & 0 \\ A_{21} & A_{22} \end{bmatrix}}_{A} \begin{bmatrix} \widetilde{V}_w(t) \\ \Delta P(t) \end{bmatrix} + \underbrace{\begin{bmatrix} -A_{11} \\ 0 \end{bmatrix}}_{B} w(t), \tag{5.9}$$

where $w(t) = (w_1(t), \ldots, w_N(t)) \in \mathbb{R}^N$ is the white Gaussian noise with covariance matrix W, $\widetilde{V}_w(t) = (\widetilde{V}_{w,1}(t), \ldots, \widetilde{V}_{w,N}(t)) \in \mathbb{R}^N$ and $\Delta P(t) = (\Delta P_1(t), \ldots, \Delta P_N(t)) \in \mathbb{R}^N$ are the states, $A_{11} = \operatorname{diag}(-\frac{1}{\tau_{w,1}}, \ldots, -\frac{1}{\tau_{w,N}})$, $A_{21} = \operatorname{diag}(\gamma_1, \ldots, \gamma_N)$,

$A_{22} = \mathrm{diag}(-\frac{1+K_{p,1}}{\tau_1}, \ldots, -\frac{1+K_{p,N}}{\tau_N})$, and $A \in \mathbb{R}^{2N\times 2N}$ and $B \in \mathbb{R}^{2N\times N}$ are as defined in (5.9), with A being asymptotically stable according to Proposition 4.2.

In Chap. 4, the gains $K_{p,i}$'s in A_{22} of (5.9) are chosen to minimize a cost function J_2, defined as

$$J_2 = \lim_{t\to\infty} \mathrm{E}\left\{\left(\sum_{i=1}^{N} \Delta P_i(t)\right)^2 + \sum_{i=1}^{N} \epsilon_i \Delta P_{d,i}^2(t)\right\}, \tag{5.10}$$

where the ϵ_i's are positive weighting factors. This cost function contains two terms, in which the first is the steady-state variance of the regulation error reflecting the smoothness of the wind farm power output, and the second is a weighted sum of the steady-state variances of the control magnitudes reflecting the control effort (which may account for turbine fatigue). Thus, finding the $K_{p,i}$'s that minimize J_2 subject to the dynamics (5.9) may be viewed as optimizing a weighted combination of the smoothness and control effort, where correlation in the wind speeds is explicitly accounted for through the covariance matrix W. Due to the linear-quadratic structure of (5.7), (5.9), and (5.10), J_2 can be expressed as

$$J_2 = \lim_{t\to\infty} \mathrm{E}\left\{ \begin{bmatrix}\widetilde{V}_w^T(t) & \Delta P^T(t)\end{bmatrix} \underbrace{\begin{bmatrix} 0 & 0 \\ 0 & Q_{22}\end{bmatrix}}_{Q} \begin{bmatrix}\widetilde{V}_w(t) \\ \Delta P(t)\end{bmatrix}\right\} = \mathrm{trace}(SQ), \tag{5.11}$$

where $Q_{22} = \mathbb{1} \cdot \mathbb{1}^T + \mathrm{diag}(\epsilon_1 K_{p,1}^2, \ldots, \epsilon_N K_{p,N}^2)$, $\mathbb{1} \in \mathbb{R}^N$ is the all-one column vector, $Q \in \mathbb{R}^{2N\times 2N}$ is as defined in (5.11), and $S = S^T \geq 0$ is the unique solution of the Lyapunov equation

$$AS + SA^T + BWB^T = 0. \tag{5.12}$$

Hence, by solving the (non-convex) optimization problem of minimizing J_2 in (5.11) and (5.12) subject to $K_{p,i} > 0\ \forall i \in \{1, 2, \ldots, N\}$, the $K_{p,i}$'s that optimize the WFC performance may be determined.

5.2.4 Problem Statement

The above method for designing the gains $K_{p,i}$'s of the WFC is based on the assumption that system (5.8) is linear, devoid of any saturation. While this assumption is valid in a neighborhood of the origin $\Delta P(t) = 0$ of the state space, the neighborhood can be rather small. Consequently, the value of J_2 as computed using (5.11) and (5.12) may be notably different from its true value as defined by (5.10). Motivated by these considerations, Sect. 5.3 addresses the problem of developing a method for calculating the $K_{p,i}$'s, which accounts for the multiple asymmetric saturations, while Sect. 5.4 compares the method with the above linear method.

5.3 QLC-Based Wind Farm Controller Design

In this section, we apply *Quasilinear Control* (QLC) theory to the wind farm control problem stated in Sect. 5.2.4. As mentioned earlier, QLC is a set of methods for designing linear controllers for systems with linear plants and nonlinear actuators and sensors. Such systems are referred to as *linear plant/nonlinear instrumentation* (LPNI). The approach of QLC is based on the method of stochastic linearization, which reduces nonlinear systems to *quasilinear* ones, with each static nonlinearity represented by an equivalent gain and an equivalent bias. Unlike the usual, Jacobian linearization, stochastic linearization is global; the price to pay is that the equivalent gains and biases depend not only on the operating point, but also on all functional blocks and exogenous signals of the system.

In [29, 85], it is shown that if the plant has sufficiently slow dynamics as compared with the bandwidth of the exogenous signals, then the quasilinear system provides a faithful estimate of the mean and variance of the output of the original LPNI system. QLC theory takes advantage of this fact and provides methods for designing controllers for LPNI systems. These methods are based on standard techniques from linear control theory (e.g., root locus, LQR, etc.), modified appropriately to account for the equivalent gains and biases. More details on QLC and the associated MATLAB-based QLC Toolbox can be found in [56].

To apply QLC to the problem, let

$$\begin{aligned} u_i(t) &\triangleq \Delta P_{d,i}(t), \\ y_i(t) &\triangleq \Delta P_i(t), \\ b_i &\triangleq \alpha_i \overline{V}^3_{w,i} - P_i^*, \\ a_i &\triangleq -P_i^*. \end{aligned}$$

With these notations, WTCS model (5.8) becomes

$$\dot{y}_i(t) = -\frac{1}{\tau_i} y_i(t) + \frac{1}{\tau_i} \operatorname{sat}_{a_i}^{b_i}(u_i(t)) + \gamma_i \widetilde{V}_{w,i}(t). \tag{5.13}$$

With $u_i(t)$ given by (5.7) and $\widetilde{V}_{w,i}(t)$ given by (5.2), the block diagram of system (5.13) is shown in Fig. 5.2a. Application of stochastic linearization to this LPNI system (see [85] for details) yields the quasilinear system shown in Fig. 5.2b, where the saturation block shaded in gray has been replaced by an equivalent gain N_i and an equivalent bias m_i, also shaded in gray. Note that since the WTCSs are decoupled, stochastic linearization of each WTCS is independent of the others. Also, it can

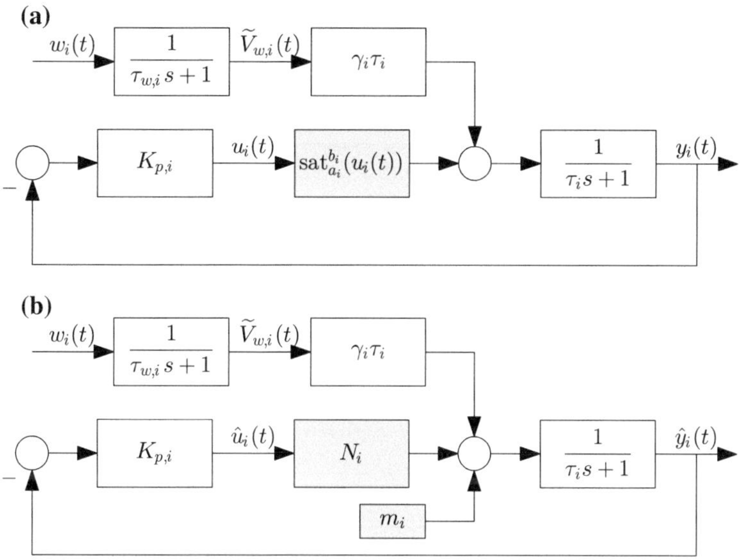

Fig. 5.2 Stochastic linearization of the LPNI system (5.13) yields a quasilinear system. LPNI system (**a**). Quasilinear system (**b**)

be shown [85] that for each $i \in \{1, 2, \ldots, N\}$, N_i and m_i may be computed by first solving the following two transcendental equations for the two unknowns N_i and M_i:

$$N_i = \frac{1}{2}\left[\mathrm{erf}\left(\frac{b_i - \mu_{\widehat{u}_i}}{\sqrt{2}\sigma_{\widehat{u}_i}}\right) - \mathrm{erf}\left(\frac{a_i - \mu_{\widehat{u}_i}}{\sqrt{2}\sigma_{\widehat{u}_i}}\right)\right], \tag{5.14}$$

$$\begin{aligned} M_i = {} & \frac{a_i + b_i}{2} + \frac{\mu_{\widehat{u}_i} - b_i}{2}\,\mathrm{erf}\left(\frac{b_i - \mu_{\widehat{u}_i}}{\sqrt{2}\sigma_{\widehat{u}_i}}\right) - \frac{\mu_{\widehat{u}_i} - a_i}{2}\,\mathrm{erf}\left(\frac{a_i - \mu_{\widehat{u}_i}}{\sqrt{2}\sigma_{\widehat{u}_i}}\right) \\ & - \frac{\sigma_{\widehat{u}_i}}{\sqrt{2\pi}}\left[\exp\left(-\left(\frac{b_i - \mu_{\widehat{u}_i}}{\sqrt{2}\sigma_{\widehat{u}_i}}\right)^2\right) - \exp\left(-\left(\frac{a_i - \mu_{\widehat{u}_i}}{\sqrt{2}\sigma_{\widehat{u}_i}}\right)^2\right)\right], \end{aligned} \tag{5.15}$$

where $\mathrm{erf}(\cdot)$ is the error function defined by

$$\mathrm{erf}(x) = \frac{2}{\sqrt{\pi}}\int_0^x e^{-t^2}\,dt,$$

and $\mu_{\widehat{u}_i}$ and $\sigma_{\widehat{u}_i}$ are, respectively, the expected value and standard deviation of the signal $\widehat{u}_i(t)$ in Fig. 5.2b given by

$$\mu_{\widehat{u}_i} = -K_{p,i} M_i,$$
$$\sigma_{\widehat{u}_i} = K_{p,i}\gamma_i\tau_i\sqrt{\frac{W_{ii}}{2(1+K_{p,i}N_i)(\tau_i+\tau_{w,i}(1+K_{p,i}N_i))}}. \tag{5.16}$$

Once the solutions N_i and M_i of (5.14) and (5.15) are found, m_i can be calculated as

$$m_i = M_i(1 + N_i K_{p,i}).$$

Furthermore, it can be shown [85] that the expected value of the output $\widehat{y}_i(t)$ in Fig. 5.2b is given by

$$\mu_{\widehat{y}_i} = M_i.$$

Finally, as far as the existence of a solution is concerned, it can be shown [85] that Eqs. (5.14) and (5.15) have a solution whenever $K_{p,i} > 0\ \forall i \in \{1, 2, \ldots, N\}$.

To reformulate the optimization problem in Sect. 5.2.3 in terms of the quasilinear system, let $y_{0,i}(t)$ denote the zero-mean part of $\widehat{y}_i(t)$, i.e.,

$$y_{0,i}(t) = \widehat{y}_i(t) - M_i.$$

Then, the cost function J_2 in (5.10) becomes

$$\begin{aligned} J_2 &= \lim_{t\to\infty} \mathrm{E}\left\{\left(\sum_{i=1}^{N}(y_{0,i}(t)+M_i)\right)^2 + \sum_{i=1}^{N}\epsilon_i K_{p,i}^2(y_{0,i}(t)+M_i)^2\right\} \\ &= \lim_{t\to\infty} \mathrm{E}\left\{ \begin{bmatrix}\widetilde{V}_w^T(t) & y_0^T(t)\end{bmatrix} \underbrace{\begin{bmatrix}0 & 0\\ 0 & Q_{22}\end{bmatrix}}_{Q} \begin{bmatrix}\widetilde{V}_w(t)\\ y_0(t)\end{bmatrix}\right\} + M^T Q_{22} M, \end{aligned} \tag{5.17}$$

where $y_0(t) = (y_{0,1}(t), \ldots, y_{0,N}(t)) \in \mathbb{R}^N$, $M = (M_1, \ldots, M_N) \in \mathbb{R}^N$, and Q_{22} and Q are as in (5.11). To express (5.17) in a form convenient for analytical calculations, notice that each $\Delta P_i(t)$ in (5.9) is the output of the system in Fig. 5.2b when $N_i = 1$ and $m_i = 0$. Likewise, each $y_{0,i}(t)$ is the output of the same system when $m_i = 0$ but N_i need not be equal to 1. Thus, by replacing its $K_{p,i}$ by $N_i K_{p,i}$, the system that produces $\Delta P_i(t)$ becomes the one that produces $y_{0,i}(t)$. This, along with the fact that $\Delta P(t)$ appears in (5.11) in the same way as $y_0(t)$ does in (5.17), implies that J_2 in (5.17) can be expressed as

$$J_2 = \mathrm{trace}(SQ) + M^T Q_{22} M, \tag{5.18}$$

where $S = S^T \geq 0$ is the unique solution of (5.12) with every $K_{p,i}$ in A_{22} replaced by $N_i K_{p,i}$. Observe that J_2 in (5.18) is a function of $3N$ variables $K_{p,i}$'s, N_i's, and M_i's, which must satisfy $2N$ equality constraints (5.14)–(5.16) and N inequality constraints $K_{p,i} > 0\ \forall i \in \{1, 2, \ldots, N\}$. Therefore, to optimize the WFC performance, we choose the $K_{p,i}$'s as the solution of the following (non-convex) optimization problem:

$$\begin{aligned} &\underset{\substack{K_{p,i}, N_i, M_i \\ i \in \{1,2,\ldots,N\}}}{\text{minimize}} && J_2 \\ &\text{subject to} && \text{(5.14)–(5.16) and } K_{p,i} > 0\ \forall i \in \{1, 2, \ldots, N\}. \end{aligned} \tag{5.19}$$

We refer to this method of designing the $K_{p,i}$'s as the QLC method and evaluate its effectiveness next.

5.4 Performance Evaluation

In this section, we evaluate analytically and via simulation the performance of WFCs designed using the linear method of Sect. 5.2.3 and the QLC method of Sect. 5.3.

5.4.1 Evaluation Scenarios

To carry out the evaluation, we divide the system parameters into two groups. The first group contains parameters to be held constant in the evaluation. These parameters and their fixed values are: the number of WTCSs $N = 10$ and, for each $i \in \{1, 2, \ldots, N\}$, the wind speed model's time constant $\tau_{w,i} = 1$, WTCS model's time constant $\tau_i = 60$, unit conversion factor $\alpha_i = 0.657$, and scalar gain $\gamma_i = 0.02$. The second group contains parameters to be varied. These parameters represent operating regimes of the WFCS and, hence, varying their values allows us to examine the WFCS performance in different regimes. For simplicity, we let their values be governed by four scalar parameters $(\mathsf{v}, \mathsf{p}, \mathsf{r}, \mathsf{e})$ in the following manner:

- *Wind speed* v: For all i and t, let the slow wind speed component $\overline{V}_{w,i}(t) = \mathsf{v}$, where $\mathsf{v} \in \{0.4, 1\}$, so that $\mathsf{v} = 0.4$ represents a low wind speed regime and $\mathsf{v} = 1$ a high. With this v, the saturation in (5.8) becomes $\text{sat}_{-P_i^*}^{0.657\mathsf{v}^3 - P_i^*}$. Thus, its linear region is narrow if $\mathsf{v} = 0.4$ and wide if $\mathsf{v} = 1$.
- *Power generation* p: For all i and t, let the intermediate power reference $P_i^*(t) = \mathsf{p}\alpha_i \overline{V}_{w,i}^3(t) = 0.657\mathsf{p}\mathsf{v}^3$, where $\mathsf{p} \in \{0.6, 1\}$. Since $\alpha_i \overline{V}_{w,i}^3(t)$ is the maximum power turbine i can generate, p is a fraction of the maximum power. Hence, $\mathsf{p} = 0.6$ represents a medium (60 %) power generation regime and $\mathsf{p} = 1$ a high (100 %). With both v and p, the saturation in (5.8) further becomes $\text{sat}_{-0.657\mathsf{p}\mathsf{v}^3}^{0.657(1-\mathsf{p})\mathsf{v}^3}$.

Therefore, its linear region is nearly symmetric about the origin if $\mathsf{p} = 0.6$ and one-sided if $\mathsf{p} = 1$.

- *Wind correlation* r: Let the covariance matrix $W = [W_{ij}]$ with $W_{ij} = \frac{1}{9}\mathsf{r}^{|i-j|}\mathsf{v}^2$, where $\mathsf{r} \in \{0, 0.999\}$, so that $\mathsf{r} = 0$ and $\mathsf{r} = 0.999$ represent, respectively, weak and strong wind correlation regimes.
- *Control penalty* e: For all i, let the control penalty $\epsilon_i = \mathsf{e}$, where $\mathsf{e} \in \{0.05, 0.1, 0.2, 0.5, 1, 2, 5, 10, 20, 100\}$, so that $\mathsf{e} = 0.05$ represents a cheap control regime and $\mathsf{e} = 100$ an expensive one.

Observe that v, p, and r each has two possible values, while e has ten. Thus, these parameters define $2^3 \cdot 10 = 80$ possible scenarios in the evaluation. For each scenario $(\mathsf{v}, \mathsf{p}, \mathsf{r}, \mathsf{e})$, we evaluate analytically and via simulation the linear and QLC methods by performing the following steps:

S1. *Linear method, analytical*: For the system parameters under consideration, use an optimization algorithm (e.g., MATLAB's fminsearch() function) to find the proportional controller gains $K_{p,i}$'s that minimize the cost function J_2 as defined by formulas (5.11) and (5.12) of the linear method. Record the corresponding minimum value of J_2 as $J^{\mathrm{a}}_{\mathrm{lin}}(\mathsf{v}, \mathsf{p}, \mathsf{r}, \mathsf{e})$.

S2. *Linear method, simulation*: With the $K_{p,i}$'s from Step S1, use MATLAB/Simulink to simulate the WFCS described by (5.2)–(5.4) from time $t = 0$ to time $t = 120{,}000$. Notice that (5.3) is nonlinear due to saturation. Assuming ergodicity and using (5.6), calculate the steady-state ensemble average in (5.10) (that determines the value of J_2) by calculating the time average from $t = 500$ to $t = 120{,}000$, where $t = 500$ is the warm-up period. Record the time average as $J^{\mathrm{s}}_{\mathrm{lin}}(\mathsf{v}, \mathsf{p}, \mathsf{r}, \mathsf{e})$.

S3. *QLC method, analytical*: Use an optimization algorithm to find the $K_{p,i}$'s that solve optimization problem (5.19) of the QLC method. Record the corresponding minimum value of J_2 as $J^{\mathrm{a}}_{\mathrm{QLC}}(\mathsf{v}, \mathsf{p}, \mathsf{r}, \mathsf{e})$.

S4. *QLC method, simulation*: With the $K_{p,i}$'s from Step S3, repeat Step S2 and record the time average as $J^{\mathrm{s}}_{\mathrm{QLC}}(\mathsf{v}, \mathsf{p}, \mathsf{r}, \mathsf{e})$.

Note that the difference between $J^{\mathrm{a}}_{\mathrm{lin}}(\cdot)$ and $J^{\mathrm{s}}_{\mathrm{lin}}(\cdot)$, and that between $J^{\mathrm{a}}_{\mathrm{QLC}}(\cdot)$ and $J^{\mathrm{s}}_{\mathrm{QLC}}(\cdot)$, quantify the accuracy of the linear and QLC methods, respectively. Moreover, the extent to which $J^{\mathrm{s}}_{\mathrm{QLC}}(\cdot)$ is less than $J^{\mathrm{s}}_{\mathrm{lin}}(\cdot)$ represents the improvement offered by the QLC method. For convenience, the percentage of such improvement is denoted as $\delta(\mathsf{v}, \mathsf{p}, \mathsf{r}, \mathsf{e})$ and defined as

$$\delta(\mathsf{v}, \mathsf{p}, \mathsf{r}, \mathsf{e}) = 100 \times \frac{J^{\mathrm{s}}_{\mathrm{lin}}(\mathsf{v}, \mathsf{p}, \mathsf{r}, \mathsf{e}) - J^{\mathrm{s}}_{\mathrm{QLC}}(\mathsf{v}, \mathsf{p}, \mathsf{r}, \mathsf{e})}{J^{\mathrm{s}}_{\mathrm{lin}}(\mathsf{v}, \mathsf{p}, \mathsf{r}, \mathsf{e})}. \tag{5.20}$$

5.4.2 Evaluation Results

Figures 5.3, 5.4, 5.5, 5.6, 5.7 show the evaluation results. Specifically, Fig. 5.3 considers the low wind speed regime ($\mathsf{v} = 0.4$), while Fig. 5.4 considers the high ($\mathsf{v} = 1$). Within these two figures, each subplot considers a particular combination of the power generation regime ($\mathsf{p} = 0.6$ for medium and $\mathsf{p} = 1$ for high) and wind correlation regime ($\mathsf{r} = 0$ for weak and $\mathsf{r} = 0.999$ for strong). Within each subplot, there are four curves representing $J^{\mathrm{a}}_{\mathrm{lin}}(\cdot)$, $J^{\mathrm{s}}_{\mathrm{lin}}(\cdot)$, $J^{\mathrm{a}}_{\mathrm{QLC}}(\cdot)$, and $J^{\mathrm{s}}_{\mathrm{QLC}}(\cdot)$ as functions of the control penalty, where the left end is the cheap control regime ($\mathsf{e} = 0.05$) and the

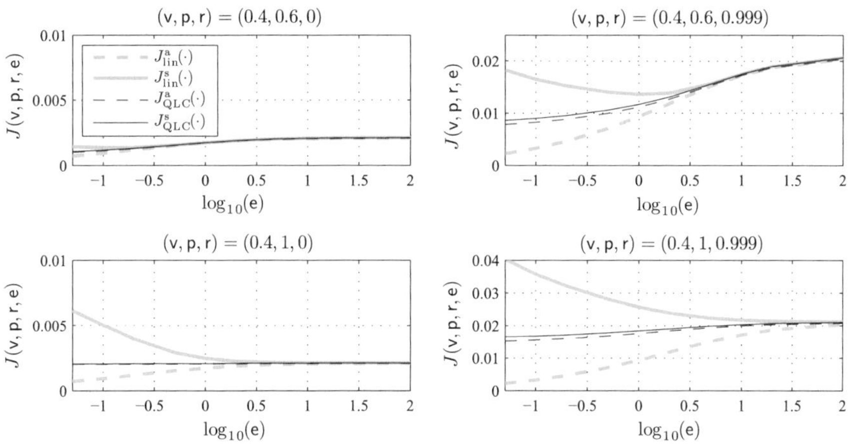

Fig. 5.3 Values of cost functions $J^{\mathrm{a}}_{\mathrm{lin}}(\cdot)$, $J^{\mathrm{s}}_{\mathrm{lin}}(\cdot)$, $J^{\mathrm{a}}_{\mathrm{QLC}}(\cdot)$, and $J^{\mathrm{s}}_{\mathrm{QLC}}(\cdot)$ in the *low* wind speed regime ($\mathsf{v} = 0.4$)

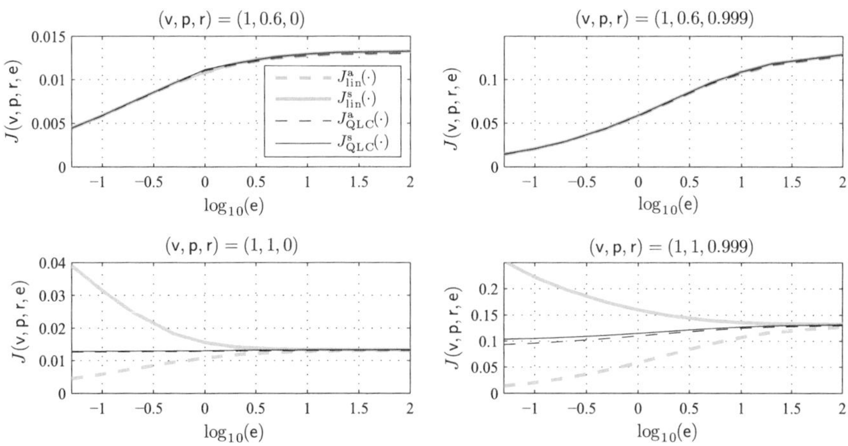

Fig. 5.4 Values of cost functions $J^{\mathrm{a}}_{\mathrm{lin}}(\cdot)$, $J^{\mathrm{s}}_{\mathrm{lin}}(\cdot)$, $J^{\mathrm{a}}_{\mathrm{QLC}}(\cdot)$, and $J^{\mathrm{s}}_{\mathrm{QLC}}(\cdot)$ in the *high* wind speed regime ($\mathsf{v} = 1$)

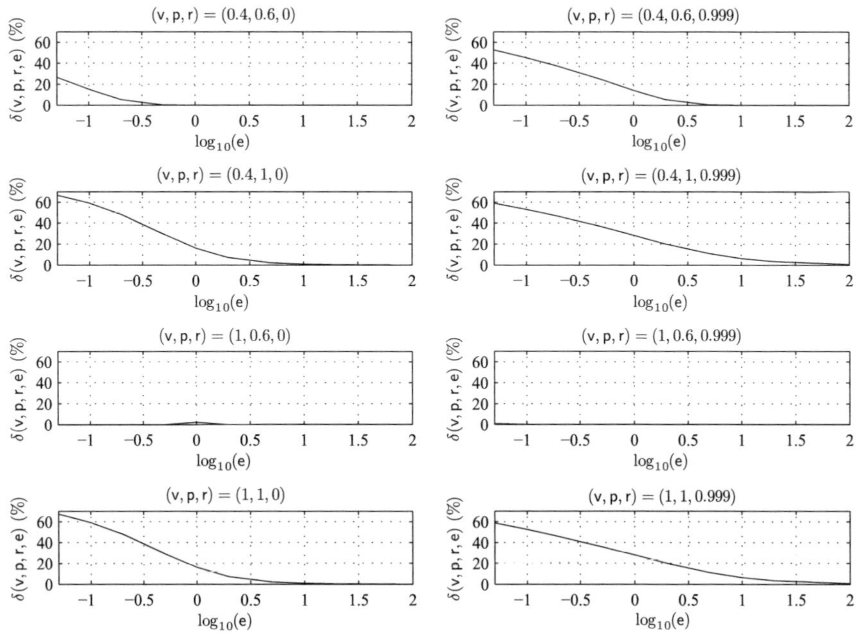

Fig. 5.5 Percentage of improvement $\delta(\mathsf{v}, \mathsf{p}, \mathsf{r}, \mathsf{e})$ across various regimes

right end is the expensive one ($\mathsf{e} = 100$). Figure 5.5 has the same format as Figs. 5.3 and 5.4 except that within each of its subplots, there is only one curve representing the percentage of improvement $\delta(\cdot)$ as a function of the control penalty, computed using (5.20). Lastly, to compare the responses of the linear and QLC methods in time domain, Figs. 5.6 and 5.7 show the simulated wind farm power output $\sum_{i=1}^{N} P_i(t)$, along with the given power reference $\sum_{i=1}^{N} P_i^*(t) = 0.657N\mathsf{p}\mathsf{v}^3$ of 6.57 and 0.420, for scenarios $(\mathsf{v}, \mathsf{p}, \mathsf{r}, \mathsf{e}) = (1, 1, 0, 0.05)$ and $(\mathsf{v}, \mathsf{p}, \mathsf{r}, \mathsf{e}) = (0.4, 1, 0, 0.05)$, respectively.

Analyzing Figs. 5.3, 5.4, 5.5, 5.6, 5.7, the following observations can be made about the linear and QLC methods:

- *Accuracy of linear method*: Regardless of the wind speed, power generation, and wind correlation regimes (i.e., $(\mathsf{v}, \mathsf{p}, \mathsf{r})$), when control is expensive (i.e., e is large), $J_{\text{lin}}^{\text{a}}(\mathsf{v}, \mathsf{p}, \mathsf{r}, \mathsf{e})$ and $J_{\text{lin}}^{\text{s}}(\mathsf{v}, \mathsf{p}, \mathsf{r}, \mathsf{e})$ are indistinguishable. This agrees with expectation because when e is large, the optimal $K_{p,i}$'s are small, causing the WFCS to operate mostly in the linear region, so that $J_{\text{lin}}^{\text{a}}(\cdot) \approx J_{\text{lin}}^{\text{s}}(\cdot)$. As control becomes cheap (i.e., e goes to zero), $J_{\text{lin}}^{\text{a}}(\mathsf{v}, \mathsf{p}, \mathsf{r}, \mathsf{e})$ approaches zero. This is also expected as it is well known that for minimum-phase linear systems, cheap control can yield arbitrarily good disturbance rejection [67]. Unexpectedly, however, as e goes to zero, not only is $J_{\text{lin}}^{\text{s}}(\mathsf{v}, \mathsf{p}, \mathsf{r}, \mathsf{e})$ bounded away from zero, it actually increases substantially in most cases. This is somewhat surprising and suggests that the linear method has poor accuracy when e is small. The result also implies that

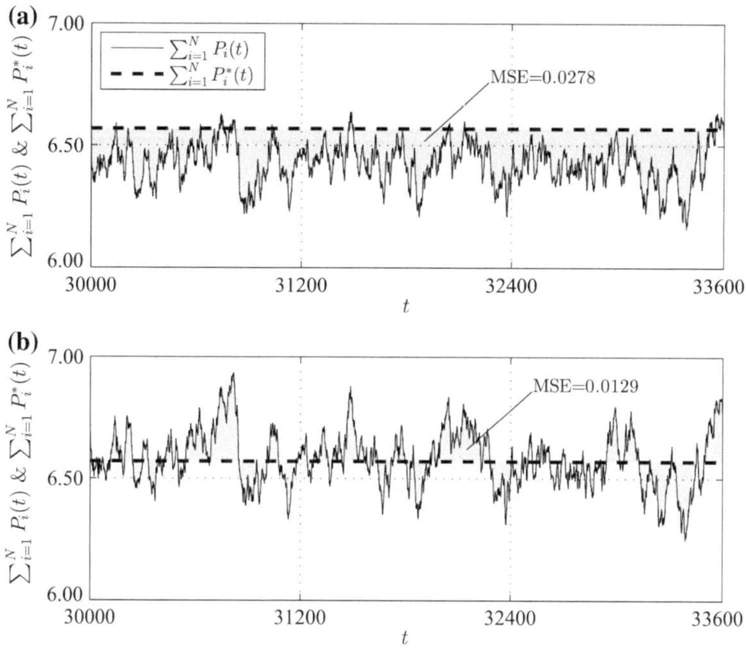

Fig. 5.6 Wind farm power output $\sum_{i=1}^{N} P_i(t)$ and power reference $\sum_{i=1}^{N} P_i^*(t)$ for scenario $(\mathsf{v}, \mathsf{p}, \mathsf{r}, \mathsf{e}) = (1, 1, 0, 0.05)$. Linear method **(a)**. QLC method **(b)**

ignoring saturation and attempting a cheap control design of the WFCS may not be advisable.

- *Accuracy of QLC method*: Similar to $J_{\text{lin}}^{\text{a}}(\cdot)$ and $J_{\text{lin}}^{\text{s}}(\cdot)$ above, regardless of $(\mathsf{v}, \mathsf{p}, \mathsf{r})$, when e is large, $J_{\text{QLC}}^{\text{a}}(\cdot)$ and $J_{\text{QLC}}^{\text{s}}(\cdot)$ are indistinguishable, which again agrees with expectation. However, unlike $J_{\text{lin}}^{\text{a}}(\cdot)$ and $J_{\text{lin}}^{\text{s}}(\cdot)$ above, as e goes to zero, $J_{\text{QLC}}^{\text{a}}(\cdot)$ and $J_{\text{QLC}}^{\text{s}}(\cdot)$ remain close to each other, with the former being slightly below the latter. This implies that the QLC method is accurate, providing an analytical means for estimating the true performance that is only slightly more optimistic than reality. The result also implies that stochastic linearization performs well for the WFCS, which has good low-pass filtering characteristics.
- *Linear method versus QLC method*: To compare the effectiveness of the two methods, consider their true performance curves $J_{\text{lin}}^{\text{s}}(\cdot)$ and $J_{\text{QLC}}^{\text{s}}(\cdot)$ in Figs. 5.3 and 5.4 and the resulting improvement curve $\delta(\cdot)$ in Fig. 5.5. Notice that no matter the operating regime $(\mathsf{v}, \mathsf{p}, \mathsf{r}, \mathsf{e})$, $J_{\text{QLC}}^{\text{s}}(\cdot)$ is never larger than $J_{\text{lin}}^{\text{s}}(\cdot)$, so that $\delta(\cdot)$ in (5.20) is always nonnegative. Moreover, as e decreases, the gap between $J_{\text{QLC}}^{\text{s}}(\cdot)$ and $J_{\text{lin}}^{\text{s}}(\cdot)$ often increases. Likewise, as e decreases, the percentage of improvement $\delta(\cdot)$ increases in most regimes, reaching 60 % in some. This result shows that the QLC method is significantly better than the linear method if control is not expensive, and as good as the linear method otherwise.

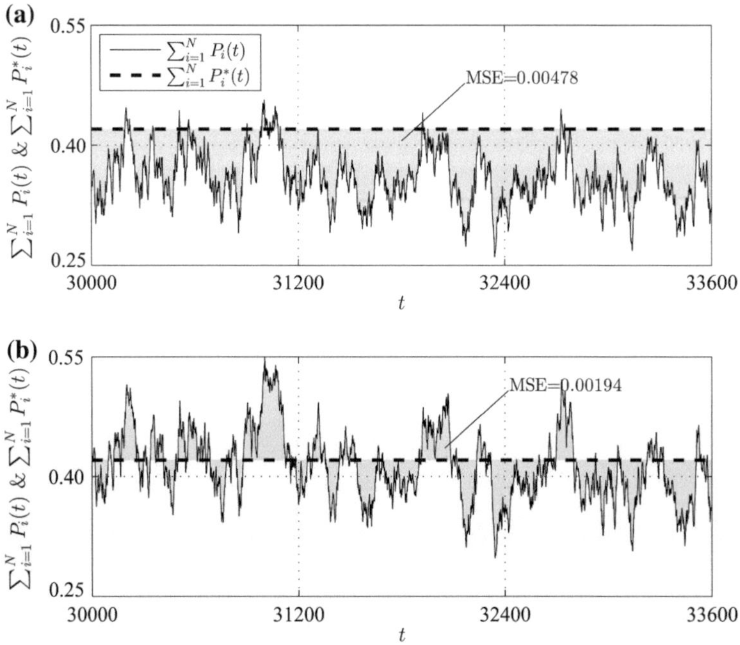

Fig. 5.7 Wind farm power output $\sum_{i=1}^N P_i(t)$ and power reference $\sum_{i=1}^N P_i^*(t)$ for scenario $(\mathsf{v}, \mathsf{p}, \mathsf{r}, \mathsf{e}) = (0.4, 1, 0, 0.05)$. Linear method **(a)**. QLC method **(b)**

- *Time-domain simulation*: Observe from Figs. 5.6 and 5.7 that the QLC method enables the wind farm power output $\sum_{i=1}^N P_i(t)$ to more accurately track the power reference $\sum_{i=1}^N P_i^*(t)$ than the linear method. Indeed, the QLC method yields mean-square errors

$$\frac{1}{3600}\int_{30000}^{33600}\left(\sum_{i=1}^{N} P_i(t) - \sum_{i=1}^{N} P_i^*(t)\right)^2 dt$$

of 0.0129 and 0.00194 that are substantially smaller than 0.0278 and 0.00478 of the linear method. These improvements may be attributed to the QLC method taking saturation into account when designing the control gains.

Figures 5.3, 5.4, 5.5 can also be used to address the following questions:

- *How do the saturation limits affect improvement?* Recall from Sect. 5.4.1 that the saturation in (5.8) has limits that depend on both v and p. In particular, it has a linear region that is narrow and nearly symmetric if $(\mathsf{v}, \mathsf{p}) = (0.4, 0.6)$, narrow and one-sided if $(\mathsf{v}, \mathsf{p}) = (0.4, 1)$, wide and nearly symmetric if $(\mathsf{v}, \mathsf{p}) = (1, 0.6)$, and wide and one-sided if $(\mathsf{v}, \mathsf{p}) = (1, 1)$. It follows from Fig. 5.5 that a linear region that is wide and one-sided yields the most improvement, one that is narrow—be it nearly symmetric or one-sided—yields the second most, and one that is wide and nearly symmetric yields the least. In terms of operating regimes, this means that the QLC method gives the most improvement in the high wind speed, high power generation regime, followed by the low wind speed regime, and subsequently by the high wind speed, medium power generation regime.

- *How does the wind correlation affect performance?* Observe that every curve within every subplot on the left-hand side of Figs. 5.3 and 5.4 is well below its counterpart on the right-hand side. This suggests that, all else being equal, the WFCS performance is notably better if the wind correlation is weak ($\mathsf{r} = 0$), compared to if it is strong ($\mathsf{r} = 0.999$). In other words, diversity in the wind speeds helps enhance the WFCS performance.

5.5 Concluding Remarks

In this chapter, we have applied the recently-developed theory of QLC to the design of proportional controller gains on the inner feedback loop of the WFC from Chap. 4. By taking the asymmetric saturation nonlinearities directly into account at the design stage, QLC leads to controller gains that are superior in performance compared to those designed by traditional linear methods which neglect saturation. Indeed, the performance improvements are found to be significant—up to 60 % reduction in the values of the cost function—over a broad range of wind farm operating regimes, including low and high wind speed, medium and high power generation, weak and strong wind correlation, and cheap and expensive control regimes.

Chapter 6
Achievability of Kinetic Energy Release in Wind Farm Active Power Control

Abstract The frontier of research in wind farm active power control is moving toward exploring the potential of *wind electric power systems* (WEPSs) in realizing the full range of functionalities offered by conventional synchronous generators. In particular, much recent attention has been given to the topic of controlling the kinetic energy stored in WEPSs as a means to maintain transient frequency stability during a power imbalance. In this chapter, we look into this topic from a theoretical standpoint and ask the fundamental question of what can, and cannot, be achieved with kinetic energy release. We first show that the classic notion of wind farm capacity, conceived at a time when WEPSs always operated in the maximum power tracking mode, is inadequate to address this question. To overcome this issue, we introduce a new concept called *achievability*, which indicates whether a given quadruplet of initial state, final time, final state, and desired wind farm power output is achievable by releasing or storing kinetic energy in the wind farm. To promote this concept of achievability, we also discuss its practical implications, present an analytical method and a numerical method for characterizing it, and illustrate the numerical method via two examples.

6.1 Introduction

The sustainability of wind power [5] is gaining increasing attention as thousands of megawatts of *wind electric power systems* (WEPSs) are built and connected to power grids in recent years. With such a tremendous progress, people begin to realize that an effective measure of wind power development is not how many WEPSs are built and installed, but how much wind power is actually utilized by consumers, and how many fossil fuel-based power plants are actually replaced. The sustainability of wind power here refers to how well the electricity generated from wind can be reliably and smoothly integrated into a modern power grid, which is a complex network where all its power generation units must work in harmony to serve the varying and uncertain demand of power. Thus, the goal of a sustainable wind power development should include its integration into the grid without compromising grid reliability, and a smooth coordination of the WEPSs with other energy sources.

J.N. Jiang et al., *Control and Operation of Grid-Connected Wind Farms*, Advances in Industrial Control, DOI 10.1007/978-3-319-39135-9_6

To accomplish this goal, a central issue that needs to be addressed is the issue of active power control for a wind farm with numerous geographically dispersed WEPSs [37]. With the rapid growth of wind power, this issue is becoming increasingly important to the power industry. The essence of active power control for a wind farm is to realize its full range of adjustability akin to what a conventional synchronous generator offers, such as load following, regulation, primary and secondary controls, as well as *maximum power tracking* (MPT). Indeed, active power control is a key enabler that allows wind farms to better serve the load and satisfy the rigid frequency stability conditions. Unfortunately however, traditional MPT-based active power control provides little contribution to frequency stability. Hence, this burden falls mostly on conventional synchronous generation. As a result, a high penetration of wind power in future generation portfolios poses a serious threat to grid reliability.

In Chaps. 2–5, we introduce several ideas and techniques for active power control along with a number of wind turbine/farm controllers. We show that these controllers enable, for example, Type 3 WEPSs (i.e., DFIGs) to operate in either the MPT mode to harvest as much wind energy as possible, or the power regulation mode to convert an appropriate amount of wind energy that depends on the need of the grid. The ability to flexibly adjust the amount of energy converted from wind, in turn, allows the WEPSs to accurately follow manual or automated instructions for frequency control. Therefore, these controllers are useful in maintaining the dynamic and steady-state stability of power grids, as well as in facilitating a smooth coordination of the WEPSs with other energy sources.

Although the controllers in Chaps. 2–5 help enhance the sustainability of wind power, they do not account for an important reliability issue, namely, the transient frequency stability, which involves the control of kinetic energy stored in moving components of WEPSs including their rotors, blades, and other moving parts. In general, frequency stability is maintained through a coordinated process of balancing the power generation and load. Moreover, the response of frequency following a significant power imbalance can typically be divided into three successive time periods—an *arresting period*, a *rebound period*, and a *recovery period*—as illustrated in Fig. 6.1.

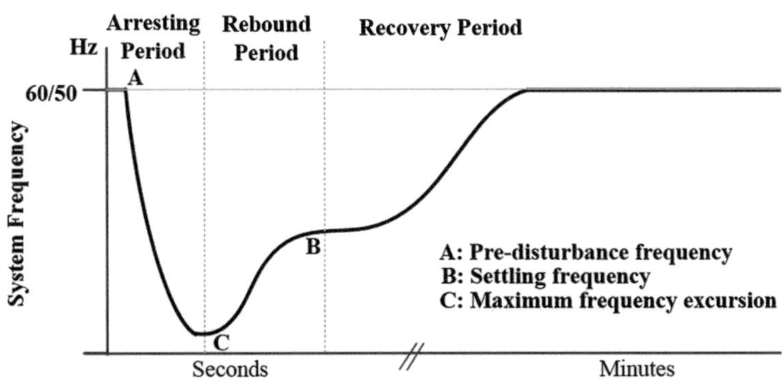

Fig. 6.1 Frequency response following a power imbalance

During the arresting period, which may be as short as a fraction of a second, it is mainly the system inertia that arrests the frequency drop to avoid triggering the special protective schemes, which may lead to cascading events. The system inertia contributes to frequency stability in two ways. First, for a small disturbance, the inertia allows the generators to stabilize the angular velocities of their rotating masses. For a synchronous generator, the angular velocity of its rotating mass is locked to the grid frequency via an electromagnetic link similar to a mechanical link. Furthermore, the larger the inertia, the lesser the angular velocity would change during the disturbance. Second, for a larger disturbance, the inertia allows kinetic energy stored in the rotating masses of the generators to be immediately released, so that the frequency drop is arrested before any primary frequency control is able to react. During the rebound period, which typically begins in a fraction of a second and ends in a few minutes, it is the primary frequency control and the droop governor that minimize the frequency deviation. Finally, during the recovery period, which usually lasts for dozens of minutes, it is the primary and secondary frequency controls that bring the frequency back to normal.

As indicated above, frequency stability is supported on different timescales by different controllers with different ranges of power. These different levels of frequency support are summarized in Table 6.1. Specifically, for the rebound and recovery periods which correspond respectively to Levels 2 and 3 in Table 6.1, the power generally comes from the prime movers, whereas for the arresting period which corresponds to Level 1, the power mostly comes from the system inertia in the form of kinetic energy released. Thus, the power from the system inertia during the arresting period may be substantially larger than that from the prime movers. Consequently, a large inertia can lower the initial *rate of change of frequency* (ROCOF) that is often used as a trigger for the so-called *Special Protection Scheme* (SPS) or *Remedial Action Scheme* (RAS). It follows that the system inertia is critical in maintaining transient frequency stability and preventing cascading failures. Lastly, Level 4 in Table 6.1 corresponds to real-time dispatches and instructions once the frequency is completely recovered. Note that, traditionally, these frequency control functions are provided by conventional synchronous generators, as opposed to by WEPSs because of the variable-speed constant-frequency requirement and the MPT control objective.

Table 6.1 Different levels of frequency support

Level	Control actions	Time-scale	Range of power needed	Objective of control	Source of energy	Function
1	Inertial response	0-5 sec.	Large	Initial transient, frequency drop, and uncontrolled	Kinetic energy stored in rotors of synchronous generators and electric machines	Moment of inertia and transient frequency control
2	Primary control by governor	2-20 sec.	Large	Arresting frequency dip and transient frequency recovery	Energy passing from primary movers of generators	Transient frequency control
3	Secondary control by AGC	4 sec. to 3 min.	Small	Dynamically following frequency target	Energy passing from primary movers of generators	Regulation
4	Real-time market	5-20 min.	Small	Stead-state, economic-dispatch, and instructions	Energy passing from primary movers of generators	Power balancing and reserve provision

In the past several years, it has been shown that the aforementioned functions can also be provided by popular Type 3 WEPSs with active power control [6, 20, 24, 60, 83, 91, 109–111]. Indeed, a few controllers have been designed for WEPSs to perform frequency control at Levels 2 through 4 [19, 26, 98]. For frequency control at Level 1, it has been shown that a WEPS may be used to produce a synthetic inertia that partially releases its kinetic energy during the arresting period [7, 10, 11, 22, 34, 36, 46, 47, 61, 79, 80, 90, 97]. In addition, the synthetic inertia has been found to be comparable in magnitude to those of conventional synchronous generators, suggesting that it may be used to help arrest the initial frequency drop. All of these findings are valuable because they open up new ways to maintain transient frequency stability, and because they alleviate concerns about reduction in system inertia when a large number of conventional synchronous generators are replaced by WEPSs.

In this chapter, we add to the theoretical understanding of frequency control at Level 1 by addressing the fundamental question of what can, and cannot, be achieved with kinetic energy release. We first describe, in Sect. 6.2, a wind farm model armed with two simplifying assumptions. Based on this model, we then demonstrate, in Sect. 6.3, two inherent limitations of the classic notion of wind farm capacity. We also show in the section that these limitations arise because wind farm capacity is a steady-state notion that does not account for transient behavior. To overcome the limitations, we next introduce, in Sect. 6.4, a new concept called *achievability*, which may be roughly defined as follows: given an initial state $\mathbf{x}(0)$ of the wind farm dynamics, a final time T, a final state $\mathbf{x}(T)$, and a desired wind farm power output $P_{d,wf}(t)$ defined over time $t \in [0, T]$, the quadruplet $(\mathbf{x}(0), T, \mathbf{x}(T), P_{d,wf})$ is said to be *achievable* if the wind farm could produce a power output that is no less than $P_{d,wf}(t)$ for all time $t \in [0, T]$ by possibly releasing or storing kinetic energy. We also explain in the section why this concept is a key to fully understand the power generation capability of a wind farm, and why it is useful in the design and analysis of practical controllers for active power control and grid frequency support. To promote the concept, we subsequently present, in Sect. 6.5, an analytical method and a numerical method for characterizing achievability. We then illustrate, in Sect. 6.6, the numerical method via a couple of examples. Finally, in Sect. 6.7 we conclude the chapter with some remarks and list some possible future research.

6.2 Wind Farm Model

Consider a wind farm consisting of $N \geq 1$ wind turbines (or WEPSs). Suppose for each wind turbine $i \in \{1, 2, \ldots, N\}$, the dynamics of its mechanical part are modeled by

$$J_i \dot{\omega}_{r,i}(t) = T_{m,i}(t) - T_{e,i}(t), \tag{6.1}$$

where $t \geq 0$ denotes time, $J_i > 0$ is the moment of inertia of its rotor, $\omega_{r,i}(t) \geq 0$ is its rotor angular velocity, $T_{m,i}(t) \in \mathbb{R}$ is its mechanical torque, $T_{e,i}(t) \in \mathbb{R}$ is its electromagnetic torque, and $T_{e,i}(t) > 0$ (respectively, $T_{e,i}(t) < 0$) means that it is acting as a generator (respectively, motor). Moreover, suppose the electrical power $P_i(t) \in \mathbb{R}$ generated by each wind turbine i is given by

$$P_i(t) = T_{e,i}(t)\omega_{r,i}(t), \tag{6.2}$$

while the mechanical power $P_{m,i}(t) \in \mathbb{R}$ captured by it is given by

$$P_{m,i}(t) = T_{m,i}(t)\omega_{r,i}(t) = \frac{1}{2}\rho A_i C_{p,i}(\lambda_i(t), \beta_i(t)) V_{w,i}^3(t), \tag{6.3}$$

where $\rho > 0$ is the air density, $A_i = \pi R_i^2$ is the area swept by its rotor blades of radius $R_i > 0$, $C_{p,i}(\cdot,\cdot) \in \mathbb{R}$ is its performance coefficient, $\beta_i(t) \in [\beta_i^{\min}, \beta_i^{\max}]$ is its blade pitch angle, $V_{w,i}(t) > 0$ is the wind speed it experiences, and $\lambda_i(t) \geq 0$ is its tip speed ratio given by

$$\lambda_i(t) = \frac{R_i \omega_{r,i}(t)}{V_{w,i}(t)}. \tag{6.4}$$

Furthermore, suppose the total electrical power $P_{wf}(t) \in \mathbb{R}$ generated by the wind farm is given by

$$P_{wf}(t) = \sum_{i=1}^{N} P_i(t), \tag{6.5}$$

which is required to match or exceed a desired power $P_{d,wf}(t) > 0$ specified by a grid operator, i.e.,

$$P_{wf}(t) \geq P_{d,wf}(t). \tag{6.6}$$

For the purpose of this study, we make two assumptions about the wind farm. First, although the blade pitch angle $\beta_i(t)$ and wind speed $V_{w,i}(t)$ of each wind turbine $i \in \{1, 2, \ldots, N\}$ are generally time-varying, we assume that they are constant, i.e.,

$$\beta_i(t) = \overline{\beta}_i, \quad V_{w,i}(t) = \overline{V}_{w,i}. \tag{6.7}$$

This assumption is not unrealistic because (6.1) has good low-pass filtering characteristics, and the timescale used in the study is short. With (6.7), the performance coefficient $C_{p,i}(\frac{R_i\omega_{r,i}(t)}{V_{w,i}(t)}, \beta_i(t))$ in (6.3) and (6.4) becomes a function of only $\omega_{r,i}(t)$ and, hence, may be written simply as $\overline{C}_{p,i}(\omega_{r,i}(t))$. Second, although the function $\overline{C}_{p,i}(\omega_{r,i}(t))$ is generally turbine-dependent [14], we assume that it is concave quadratic with a maximum of $\overline{C}_{p,i}^{\max} \in (0, 1)$ and two real roots at 0 and $\omega_{r,i}^{\max} > 0$, i.e.,

$$\overline{C}_{p,i}(\omega_{r,i}(t)) = \frac{4\overline{C}_{p,i}^{\max}}{(\omega_{r,i}^{\max})^2}\omega_{r,i}(t)(\omega_{r,i}^{\max} - \omega_{r,i}(t)). \tag{6.8}$$

This assumption is also not unreasonable because under condition (6.7) and for most values of $\overline{\beta}_i$ and $\overline{V}_{w,i}$, the performance coefficients of most wind turbines are indeed concave over a large interval of $\omega_{r,i}(t)$ and have two real roots [14]. With (6.7) and (6.8) and with a parameter $\gamma_i > 0$ defined as

$$\gamma_i = \frac{2\rho A_i \overline{C}_{p,i}^{\max} \overline{V}_{w,i}^3}{(\omega_{r,i}^{\max})^2}, \tag{6.9}$$

the nonlinear dynamics (6.1), (6.3), and (6.4) of each wind turbine i in the wind farm become

$$\dot{\omega}_{r,i}(t) = \frac{\gamma_i}{J_i}(\omega_{r,i}^{\max} - \omega_{r,i}(t)) - \frac{1}{J_i}T_{e,i}(t), \tag{6.10}$$

which is simpler as (6.10) is a first-order, affine time-invariant system with state $\omega_{r,i}(t)$ and input $T_{e,i}(t)$. Note, however, that system (6.10) is subject to a state constraint

$$\omega_{r,i}(t) \geq 0 \tag{6.11}$$

because wind turbines typically have mechanisms that prevent their rotors from turning backward.

6.3 Limitations of the Notion of Wind Farm Capacity

When we look at a wind farm, we often want to know what it is capable of, in terms of generating power. To answer this question, we often look at its *capacity*, commonly defined as *the maximum power it can generate in steady-state* (i.e., in a sustainable fashion). Although this classic notion of capacity is very useful (e.g., in power systems planning), it is not indicative of what the wind farm can produce during transient. Indeed, the wind farm could momentarily slow down or speed up its turbine rotors to release or store kinetic energy, temporarily producing power that goes way above or below its capacity. Therefore, although useful, the notion of capacity provides only a glimpse into what a wind farm is truly capable of.

To mathematically demonstrate the above point, consider the following propositions, in which Proposition 6.1 illustrates the calculation of capacity, and Propositions 6.2 and 6.3 reveal its limitations:

Proposition 6.1 (Calculation of Capacity) *The capacity of the wind farm modeled in Sect. 6.2 is given by*

$$\frac{1}{4}\sum_{i=1}^{N}\gamma_i(\omega_{r,i}^{\max})^2$$

or equivalently

$$\frac{1}{2}\rho\sum_{i=1}^{N}A_i\overline{C}_{p,i}^{\max}\overline{V}_{w,i}^3.$$

Proof The total power generated by the wind farm can be expressed using (6.5), (6.2), and (6.10) as

$$\begin{aligned}P_{wf}(t) &= \sum_{i=1}^{N}P_i(t) = \sum_{i=1}^{N}T_{e,i}(t)\omega_{r,i}(t)\\ &= \sum_{i=1}^{N}\gamma_i\omega_{r,i}(t)(\omega_{r,i}^{\max}-\omega_{r,i}(t)) - J_i\omega_{r,i}(t)\dot{\omega}_{r,i}(t). \end{aligned}\tag{6.12}$$

In steady-state, $\dot{\omega}_{r,i}(t) = 0$ for all $i \in \{1, 2, \ldots, N\}$. Hence, in steady-state each $\omega_{r,i}(t)$ in (6.12) may be replaced by a constant $\overline{\omega}_{r,i}$, so that

$$P_{wf}(t) = \sum_{i=1}^{N}\gamma_i\overline{\omega}_{r,i}(\omega_{r,i}^{\max} - \overline{\omega}_{r,i}).\tag{6.13}$$

Observe that $P_{wf}(t)$ in (6.13) is maximized when $\overline{\omega}_{r,i} = \frac{1}{2}\omega_{r,i}^{\max}$ for all $i \in \{1, 2, \ldots, N\}$, and its maximum value is $\frac{1}{4}\sum_{i=1}^{N}\gamma_i(\omega_{r,i}^{\max})^2$. Thus, the wind farm capacity is $\frac{1}{4}\sum_{i=1}^{N}\gamma_i(\omega_{r,i}^{\max})^2$, which can also be expressed using (6.9) as $\frac{1}{2}\rho\sum_{i=1}^{N}A_i\overline{C}_{p,i}^{\max}\overline{V}_{w,i}^3$. □

Proposition 6.2 (A Limitation of Capacity) *Consider the wind farm modeled in Sect. 6.2 and suppose it has only $N = 1$ wind turbine. Also suppose the grid operator wants a constant desired power $P_{d,wf}(t) = \overline{P}_{d,wf} > 0$. Then, for every $\overline{P}_{d,wf}$ above the wind farm capacity (i.e., $\overline{P}_{d,wf} > \frac{1}{4}\gamma_1(\omega_{r,1}^{\max})^2$) and every initial state $\omega_{r,1}(0) > 0$: (i) there exist $t^* \in (0,\infty)$ and an input $T_{e,1}(t) \in \mathbb{R}$ defined over $t \in [0, t^*]$ such that requirement (6.6) holds (i.e., $P_{wf}(t) \geq P_{d,wf}(t)$) for all $t \in [0, t^*]$; and (ii) there exists no input $T_{e,1}(t) \in \mathbb{R}$ defined over $t \in [0,\infty)$ such that (6.6) holds for all $t \in [0,\infty)$.*

Proof Let $\overline{P}_{d,wf} > \frac{1}{4}\gamma_1(\omega_{r,1}^{\max})^2$ and $\omega_{r,1}(0) > 0$ be given. To prove (i), consider a scalar nonlinear differential equation

$$\dot{x}(t) = \frac{\gamma_1}{J_1}(\omega_{r,1}^{\max} - x(t)) - \frac{1}{J_1}\frac{\overline{P}_{d,wf}}{x(t)} \tag{6.14}$$

with initial condition $x(0) = \omega_{r,1}(0)$. Since $x(0) > 0$, the right-hand side of (6.14) is locally Lipschitz in a small neighborhood of $x(0)$. Hence, there exists $t^* \in (0, \infty)$ such that (6.14) has a unique solution $x(t)$ over $t \in [0, t^*]$ [62]. Moreover, since $x(0) > 0$, t^* can be taken so that $x(t) > 0$ for all $t \in [0, t^*]$. Now let $T_{e,1}(t) = \frac{\overline{P}_{d,wf}}{x(t)}$ for all $t \in [0, t^*]$, which is well-defined. Then, system (6.10) becomes

$$\dot{\omega}_{r,1}(t) = \frac{\gamma_1}{J_1}(\omega_{r,1}^{\max} - \omega_{r,1}(t)) - \frac{1}{J_1}\frac{\overline{P}_{d,wf}}{x(t)}. \tag{6.15}$$

It follows from (6.14) that $\omega_{r,1}(t) = x(t)$ is the unique solution to (6.15) over $t \in [0, t^*]$. Due to (6.5), (6.2), and the above,

$$P_{wf}(t) = P_1(t) = T_{e,1}(t)\omega_{r,1}(t) = \frac{\overline{P}_{d,wf}}{x(t)}x(t) = \overline{P}_{d,wf} = P_{d,wf}(t)$$

for all $t \in [0, t^*]$. Therefore, (6.6) holds for all $t \in [0, t^*]$. To prove (ii), assume to the contrary that there exists $T_{e,1}(t) \in \mathbb{R}$ defined over $t \in [0, \infty)$ such that the solution $\omega_{r,1}(t)$ of (6.10) satisfies $P_{wf}(t) \geq P_{d,wf}(t)$ for all $t \in [0, \infty)$. Then, from (6.5) and (6.2), we obtain $T_{e,1}(t)\omega_{r,1}(t) \geq P_{d,wf}(t)$ for all $t \in [0, \infty)$. Since $P_{d,wf}(t) = \overline{P}_{d,wf} > 0$ for all $t \in [0, \infty)$ and $\omega_{r,1}(t) \geq 0$ for all $t \in [0, \infty)$ due to (6.11), we have

$$\omega_{r,1}(t) > 0 \text{ for all } t \in [0, \infty) \tag{6.16}$$

and $T_{e,1}(t) \geq \frac{\overline{P}_{d,wf}}{\omega_{r,1}(t)}$ for all $t \in [0, \infty)$. Substituting the latter into (6.10) and rearranging it yield

$$\dot{\omega}_{r,1}(t) \leq \frac{-\gamma_1\omega_{r,1}^2(t) + \gamma_1\omega_{r,1}^{\max}\omega_{r,1}(t) - \overline{P}_{d,wf}}{J_1\omega_{r,1}(t)}. \tag{6.17}$$

Notice that the numerator of the right-hand side of (6.17) is a second-order polynomial in $\omega_{r,1}(t)$ with a maximum value of $\frac{1}{4}\gamma_1(\omega_{r,1}^{\max})^2 - \overline{P}_{d,wf}$. Since $\overline{P}_{d,wf} > \frac{1}{4}\gamma_1(\omega_{r,1}^{\max})^2$, this maximum value is negative. This, along with (6.16) and (6.17), implies that $\dot{\omega}_{r,1}(t) \leq \frac{-c}{\omega_{r,1}(t)}$ for all $t \in [0, \infty)$, where $c = (\overline{P}_{d,wf} - \frac{1}{4}\gamma_1(\omega_{r,1}^{\max})^2)/J_1 > 0$. This and (6.16), in turn, imply that $\dot{\omega}_{r,1}(t) < 0$ for all $t \in [0, \infty)$, so that $0 < \omega_{r,1}(t) \leq \omega_{r,1}(0)$ for all $t \in [0, \infty)$. Thus, $\dot{\omega}_{r,1}(t) \leq \frac{-c}{\omega_{r,1}(0)}$ for all $t \in [0, \infty)$. It follows that there exists $t' \in (0, \frac{\omega_{r,1}^2(0)}{c}]$ such that $\omega_{r,1}(t') = 0$, contradicting (6.16) and proving (ii). □

Proposition 6.2 says that, in theory, no matter how far above the wind farm capacity the constant desired power is, and no matter how slowly the wind turbine rotor is initially turning, the wind farm is able to satisfy the grid operator requirement (6.6) for a finite amount of time because the turbine could rapidly release its kinetic energy. However, as long as the constant desired power is above the capacity, no matter how quickly the turbine rotor is initially turning, the wind farm is unable to satisfy requirement (6.6) indefinitely because the turbine only has a finite amount of kinetic energy to release. In fact, the proof of Proposition 6.2 implies that the turbine would lose all its kinetic energy and come to a stop at some finite time t', which depends on the constant desired power $\overline{P}_{d,wf}$ and the initial state $\omega_{r,1}(0)$. Thus, *just because the constant desired power is above the wind farm capacity it does not mean that requirement* (6.6) *cannot be met—it just cannot be sustained.* This analysis reveals an important limitation of capacity: it does not tell how much power a wind farm can generate during transient.

Proposition 6.3 (Another Limitation of Capacity) *Consider the wind farm modeled in Sect. 6.2 and suppose it has only $N = 1$ wind turbine. Also suppose the grid operator wants a constant desired power $P_{d,wf}(t) = \overline{P}_{d,wf} > 0$. Then, for every $\overline{P}_{d,wf}$ below or equal to the wind farm capacity (i.e., $\overline{P}_{d,wf} \le \frac{1}{4}\gamma_1(\omega_{r,1}^{\max})^2$), there exists $\omega_{r,1}^* \in (0,\infty)$ such that for every initial state $\omega_{r,1}(0) \in (0,\omega_{r,1}^*)$, there exists no input $T_{e,1}(t) \in \mathbb{R}$ defined over $t \in [0,\infty)$ such that (6.6) holds for all $t \in [0,\infty)$.*

Proof Let $\overline{P}_{d,wf} \le \frac{1}{4}\gamma_1(\omega_{r,1}^{\max})^2$ be given. Let

$$\omega_{r,1}^* = \frac{\omega_{r,1}^{\max}}{2} - \sqrt{\frac{\frac{1}{4}\gamma_1(\omega_{r,1}^{\max})^2 - \overline{P}_{d,wf}}{\gamma_1}}.$$

Note that $\omega_{r,1}^* \in \mathbb{R}$ and $\omega_{r,1}^* > 0$, the latter due to $\overline{P}_{d,wf} > 0$. Let $\omega_{r,1}(0) \in (0,\omega_{r,1}^*)$ be given. Assume to the contrary that there exists $T_{e,1}(t) \in \mathbb{R}$ defined over $t \in [0,\infty)$ such that the solution $\omega_{r,1}(t)$ of (6.10) satisfies $P_{wf}(t) \ge P_{d,wf}(t)$ for all $t \in [0,\infty)$. Then, using the same arguments as those in the proof of (ii) of Proposition 6.2, it can be shown that (6.16) and (6.17) hold. Since $\gamma_1 > 0$ and $\overline{P}_{d,wf} \le \frac{1}{4}\gamma_1(\omega_{r,1}^{\max})^2$, the second-order polynomial

$$f(\omega_{r,1}(t)) \triangleq -\gamma_1\omega_{r,1}^2(t) + \gamma_1\omega_{r,1}^{\max}\omega_{r,1}(t) - \overline{P}_{d,wf}$$

in (6.17) is concave with two positive real roots at $\omega_{r,1}^*$ and $\omega_{r,1}^{\max} - \omega_{r,1}^*$. Hence: (a) $f(x) < 0$ for all $x \in (0,\omega_{r,1}^*)$; and (b) $f(x) \le f(y)$ for all $x, y \in (0,\omega_{r,1}^*)$ with $x \le y$. Due to the given $\omega_{r,1}(0) \in (0,\omega_{r,1}^*)$, property (a), (6.16), and (6.17), we have $\dot{\omega}_{r,1}(t) < 0$ for all $t \in [0,\infty)$, so that $0 < \omega_{r,1}(t) \le \omega_{r,1}(0)$ for all $t \in [0,\infty)$. This and (6.17), along with properties (a) and (b), imply that

$$\dot{\omega}_{r,1}(t) \le \frac{f(\omega_{r,1}(t))}{J_1\omega_{r,1}(t)} \le \frac{f(\omega_{r,1}(0))}{J_1\omega_{r,1}(t)} \le \frac{f(\omega_{r,1}(0))}{J_1\omega_{r,1}(0)} < 0$$

for all $t \in [0,\infty)$. Therefore, there exists $t' \in (0, \frac{-J_1\omega_{r,1}^2(0)}{f(\omega_{r,1}(0))}]$ such that $\omega_{r,1}(t') = 0$, contradicting (6.16) and proving the proposition. □

Proposition 6.3 says that, theoretically, no matter how far below the wind farm capacity the constant desired power is, if the wind turbine rotor is initially turning too slowly, the wind farm is unable to satisfy the grid operator requirement (6.6) at all times because the turbine does not have enough kinetic energy to compensate for the shortage in mechanical power it could convert from wind. Hence, *just because the constant desired power is below the wind farm capacity does not mean that requirement* (6.6) *can be sustained—it depends, again, on the constant desired power* $\overline{P}_{d,wf}$ *and the initial state* $\omega_{r,1}(0)$. This analysis reveals another important limitation of capacity that stems from its inability to tell how much power a wind farm can generate during transient.

6.4 Concept of Achievability

As can be seen from Sect. 6.3, the steady-state notion of capacity does not "tell the whole story" about the capability of a wind farm in generating power. Also seen is that such capability depends on the desired power, the initial state, and the time duration of interest. Motivated by these observations, in this section we introduce a new concept called *achievability* that complements the notion of capacity by capturing *both* the steady-state and transient power generation capability of a wind farm. To begin, let

$$\begin{aligned}\omega_r(t) &= (\omega_{r,1}(t), \ldots, \omega_{r,N}(t)) \in \mathbb{R}^N,\\ \omega_r^{\max} &= (\omega_{r,1}^{\max}, \ldots, \omega_{r,N}^{\max}) \in \mathbb{R}^N,\\ T_e(t) &= (T_{e,1}(t), \ldots, T_{e,N}(t)) \in \mathbb{R}^N,\\ J &= \mathrm{diag}(J_1, \ldots, J_N) \in \mathbb{R}^{N\times N},\\ \Gamma &= \mathrm{diag}(\gamma_1, \ldots, \gamma_N) \in \mathbb{R}^{N\times N}.\end{aligned}$$

Using these vector-matrix notations as well as (6.5) and (6.2), the wind farm dynamics (6.10), state constraint (6.11), and grid operator requirement (6.6) may be compactly written as

$$\dot{\omega}_r(t) = \Gamma J^{-1}(\omega_r^{\max} - \omega_r(t)) - J^{-1}T_e(t), \tag{6.18}$$

$$\omega_r(t) \succeq 0, \tag{6.19}$$

$$P_{wf}(t) = \omega_r(t)^\top T_e(t) \geq P_{d,wf}(t), \tag{6.20}$$

where $\top$ denotes transposition and $\succeq$ denotes element-wise greater-than-or-equal-to. With (6.18)–(6.20), the concept of achievability can be readily defined as follows:

Definition 6.1 (*Achievability*) Given an initial state $\omega_r^0 \succeq 0$, a final time $T \in (0, \infty)$, a final state $\omega_r^T \succeq 0$, and a desired wind farm power output $P_{d,wf}(t) > 0$ defined over $t \in [0, T]$, if there exists an input $T_e(t) \in \mathbb{R}^N$ defined over $t \in [0, T]$ such that the state $\omega_r(t)$ satisfies $\omega_r(0) = \omega_r^0$, $\omega_r(T) = \omega_r^T$, and (6.18)–(6.20) for all $t \in [0, T]$, then the quadruplet $(\omega_r^0, T, \omega_r^T, P_{d,wf})$ is said to be *achievable*. Otherwise, it is said to be *unachievable*.

Observe that, unlike capacity which is a single number that captures only the steady-state behavior of a wind farm, achievability introduced in Definition 6.1 captures also its transient behavior by indicating whether it is possible to find an input $T_e(t)$ that drives the state $\omega_r(t)$ of the wind farm dynamics (6.18) from a given initial value ω_r^0 at time 0 to a given final value ω_r^T at a given final time T, all the while satisfying the state constraint (6.19) and causing the wind farm to produce a total power that satisfies the grid operator requirement (6.20). Thus, given a quadruplet $(\omega_r^0, T, \omega_r^T, P_{d,wf})$, it is of interest to be able to determine whether the quadruplet is achievable.

The ability to make such a determination is crucial because it allows us to understand what can and cannot be achieved with kinetic energy release. For instance, we could determine whether a wind farm could produce an additional power output of, say, 10 MW for 1 s if the state of the wind farm has a certain initial value and if we would like it to have a certain final value (perhaps getting ready for the next round of frequency support via kinetic energy release). If we determine that such a combination of specifications (i.e., the quadruplet) is unachievable, we could start asking which one of the specifications (i.e., element of the quadruplet) needs to be relaxed, and by how much. On the contrary, if we determine that such a combination of specifications is achievable, we could start asking what is the best way to achieve that so that, for instance, the mechanical movements of the turbines is minimized for fatigue reduction. Alternatively, we could ask to what extent can the actual wind farm power output $P_{wf}(t)$ be increased, or to what extent can the final time T be increased with all else being fixed, while maintaining the achievability of the quadruplet. Furthermore, given that wind farms are often deployed in conjunction with other conventional and renewable energy sources, we could extend the concept of achievability in Definition 6.1 by viewing both the wind farm and these other energy sources as a single entity. Therefore, this concept is a key to fully understand the power generation capability of a wind farm, which in turn is useful in the design and analysis of practical controllers for active power control and grid frequency support.

6.5 Characterization of Achievability

Because achievability is a useful concept, we want to have methods for characterizing it, which may be used to determine whether a quadruplet is achievable without having to search for the input as required by Definition 6.1. In more precise terms, if $\mathcal{S}$ represents the set of all quadruplets $(\omega_r^0, T, \omega_r^T, P_{d,wf})$, and $\mathcal{A} \subset \mathcal{S}$ represents the set of all quadruplets that are achievable, ideally we want to know how the set $\mathcal{A}$ looks like, preferably with little effort. In this section, we present an analytical method and a numerical method for characterizing the set $\mathcal{A}$.

6.5.1 Analytical Method

To describe the analytical method, consider the following theorem, which provides a sufficient condition for a quadruplet to not be in the set $\mathcal{A}$. Note that unlike Propositions 6.2 and 6.3, this theorem is valid for a general wind farm with an arbitrary number $N \geq 1$ of wind turbines and an arbitrary time-varying desired power $P_{d,wf}(t) > 0$:

Theorem 6.1 *Consider the wind farm modeled in Sect. 6.2. For each* $\omega_r^0 \succeq 0$, $T \in (0, \infty)$, $\omega_r^T \succeq 0$, *and function* $P_{d,wf}(t) > 0$ *defined over* $t \in [0, T]$, *if*

$$\inf_{t\in[0,T]} P_{d,wf}(t) > \frac{1}{4}(\omega_r^{\max})^\top \Gamma \omega_r^{\max} \tag{6.21}$$

and

$$T \geq \frac{\frac{1}{2}(\omega_r^0)^\top J \omega_r^0}{\inf_{t\in[0,T]} P_{d,wf}(t) - \frac{1}{4}(\omega_r^{\max})^\top \Gamma \omega_r^{\max}}, \tag{6.22}$$

then the quadruplet $(\omega_r^0, T, \omega_r^T, P_{d,wf})$ *is unachievable, i.e.,*

$$(\omega_r^0, T, \omega_r^T, P_{d,wf}) \notin \mathcal{A}.$$

Proof Let $K(t)$ denote the total kinetic energy of the N wind turbines at time t, defined as

$$K(t) = \frac{1}{2}\sum_{i=1}^{N} J_i \omega_{r,i}^2(t) = \frac{1}{2}\omega_r(t)^\top J \omega_r(t). \tag{6.23}$$

Taking the time derivative of $K(t)$ in (6.23) along the state trajectory $\omega_r(t)$ of dynamics (6.18) and using (6.20), we obtain

$$\begin{aligned} \dot{K}(t) &= \omega_r(t)^\top \Gamma(\omega_r^{\max} - \omega_r(t)) - \omega_r(t)^\top T_e(t) \\ &\leq \underbrace{\omega_r(t)^\top \Gamma(\omega_r^{\max} - \omega_r(t))}_{q(\omega_r(t))} - P_{d,wf}(t), \end{aligned} \tag{6.24}$$

where $q : \mathbb{R}^N \to \mathbb{R}$ is a concave quadratic function given by

$$q(z) = z^\top \Gamma(\omega_r^{\max} - z),$$

which has a maximizer at $z^* = \frac{1}{2}\omega_r^{\max}$ and a maximum value of

$$q(z^*) = \frac{1}{4}(\omega_r^{\max})^\top \Gamma \omega_r^{\max} > 0.$$

Hence, if (6.21) holds, $\dot{K}(t)$ in (6.24) may be written as

$$\dot{K}(t) \leq q(z^*) - P_{d,wf}(t) \leq \frac{1}{4}(\omega_r^{\max})^\top \Gamma \omega_r^{\max} - \inf_{t \in [0,T]} P_{d,wf}(t) < 0.$$

If (6.22) also holds, there exists $t' \in [0, T]$ such that $K(t') = 0$, so that $\omega_r(t') = 0$ by (6.23), which violates (6.20). Thus, $(\omega_r^0, T, \omega_r^T, P_{d,wf}) \notin \mathcal{A}$. □

Observe that conditions (6.21) and (6.22) in Theorem 6.1 represent an easy-to-check sufficient condition for a quadruplet to not be in the set $\mathcal{A}$. These two conditions therefore describe an easy-to-use analytical method for characterizing the set $\mathcal{A}$. We note that additional, possibly less conservative necessary and/or sufficient conditions for a quadruplet to be or not be in the set $\mathcal{A}$ may also be derived, which would lead to alternative, potentially more powerful analytical methods. Such derivation thus represents a viable future research direction.

6.5.2 Numerical Method

To describe the numerical method, we first discretize the continuous-time wind farm dynamics (6.18), state constraint (6.19), and grid operator requirement (6.20) in the following manner: let time $t \geq 0$ be slotted and let each time slot $k \in \{0, 1, 2, \ldots\}$ run from time kT_s to time $(k+1)T_s$, where $T_s > 0$ is the time slot duration that is meant to be small. Suppose both the input $T_e(t)$ and desired wind farm power output $P_{d,wf}(t)$ are constant in each time slot and denoted as $T_e[k]$ and $P_{d,wf}[k]$ with square brackets, i.e., $T_e(t) = T_e[k]$ and $P_{d,wf}(t) = P_{d,wf}[k]$ for all $t \in [kT_s, (k+1)T_s)$ and $k \in \{0, 1, 2, \ldots\}$. Then, we may discretize (6.18) via a zero-order hold to get

$$\omega_r[k+1] = e^{-\Gamma J^{-1} T_s}\omega_r[k] + (I - e^{-\Gamma J^{-1} T_s})(\omega_r^{\max} - \Gamma^{-1} T_e[k]), \tag{6.25}$$

and express (6.19) and (6.20) as

$$\omega_r[k+1] \succeq 0, \tag{6.26}$$

$$\omega_r[k]^\top T_e[k] \geq P_{d,wf}[k], \tag{6.27}$$

where the state $\omega_r[k]$ is the value of $\omega_r(t)$ at time $t = kT_s$, $e^{(\cdot)}$ denotes the matrix exponential, and (6.26) is written as "$\omega_r[k+1] \succeq 0$" instead of "$\omega_r[k] \succeq 0$" for convenience. Notice that because J and Γ are diagonal matrices with positive diagonal entries, so are J^{-1}, Γ^{-1}, $e^{-\Gamma J^{-1} T_s}$, and $I - e^{-\Gamma J^{-1} T_s}$ in (6.25). This property will be useful shortly.

Having obtained the discrete-time wind farm dynamics (6.25), state constraint (6.26), and grid operator requirement (6.27), we next establish the following claim: *for any given $\omega_r[k] \succeq 0$, the set of $\omega_r[k+1]$ satisfying (6.25)–(6.27), denoted as $\mathcal{P}(\omega_r[k]) \subset \mathbb{R}^N$, is either empty or a convex polytope.* Let $\omega_r[k] \succeq 0$ be given. If

$\omega_r[k] = 0$, then (6.27) cannot be satisfied since $P_{d,wf}[k] > 0$. As a result, the set $\mathcal{P}(\omega_r[k])$ is empty. If $\omega_r[k] \succeq 0$ but $\omega_r[k] \neq 0$—as illustrated in Fig. 6.2a for the case of $N = 2$—then clearly there exists $T_e[k] \in \mathbb{R}^N$ such that (6.27) holds. In fact, such $T_e[k]$ may be parameterized by

$$T_e[k] = \underbrace{\frac{P_{d,wf}[k]}{\|\omega_r[k]\|^2}\omega_r[k]}_{\eta_1} + \underbrace{\varepsilon\omega_r[k]}_{\eta_2} + \underbrace{B\xi}_{\eta_3}, \tag{6.28}$$

where $\varepsilon \geq 0$ and $\xi \in \mathbb{R}^{N-1}$ are parameters, while $B \in \mathbb{R}^{N\times(N-1)}$ is a matrix whose columns form an orthonormal basis of the nullspace of $\omega_r[k]$, i.e., $\omega_r[k]^\top B = 0$. To understand this parameterization, let $\eta_1, \eta_2, \eta_3 \in \mathbb{R}^N$ be as defined in (6.28), so that $T_e[k] = \eta_1 + \eta_2 + \eta_3$. Note that η_1 is the minimum-norm solution to $\omega_r[k]^\top T_e[k] = P_{d,wf}[k]$ and, thus, is a vector in the direction of $\omega_r[k]$ that "barely" satisfies (6.27). Likewise, η_2 is also a vector in the direction of $\omega_r[k]$ whose parameter ε represents the margin with which (6.27) is oversatisfied. Lastly, η_3 is a vector orthogonal to $\omega_r[k]$ whose parameter ξ has no impact on (6.27). The vectors $\omega_r[k], \eta_1, \eta_2, \eta_3, T_e[k]$, along with the set of $T_e[k]$ satisfying (6.27), are illustrated in Fig. 6.2b for the case of $N = 2$.

Substituting $T_e[k]$ in (6.28) into the dynamics (6.25), we obtain

$$\begin{aligned}\omega_r[k+1] = {} & \underbrace{e^{-\Gamma J^{-1}T_s}\omega_r[k] + (I - e^{-\Gamma J^{-1}T_s})\Big(\omega_r^{\max} - \Gamma^{-1}\frac{P_{d,wf}[k]}{\|\omega_r[k]\|^2}\omega_r[k]\Big)}_{\mu_1} \\ & + \underbrace{(-1)(I - e^{-\Gamma J^{-1}T_s})\Gamma^{-1}\varepsilon\omega_r[k]}_{\mu_2} \\ & + \underbrace{(-1)(I - e^{-\Gamma J^{-1}T_s})\Gamma^{-1}B\xi}_{\mu_3}.\end{aligned} \tag{6.29}$$

To understand this expression, let $\mu_1, \mu_2, \mu_3 \in \mathbb{R}^N$ be as defined in (6.29), so that $\omega_r[k+1] = \mu_1 + \mu_2 + \mu_3$. Observe that μ_1 is a fixed vector that does not depend on the parameters ε and ξ. In contrast, μ_2 and μ_3 are vectors parameterized by ε and ξ, respectively. Moreover, $\mu_2 \preceq 0$, μ_3 is orthogonal to the vector $\mu_3' \triangleq (I - e^{-\Gamma J^{-1}T_s})^{-1}\Gamma\omega_r[k] \in \mathbb{R}^N$, and $\mu_3' \succeq 0$ with $\mu_3' \neq 0$. Furthermore, $\omega_r[k+1] = \mu_1 + \mu_2 + \mu_3$ must satisfy (6.26). Hence, the set $\mathcal{P}(\omega_r[k])$ of $\omega_r[k+1]$ satisfying (6.25)–(6.27) is a convex polytope of the form

$$\mathcal{P}(\omega_r[k]) = \{x \in \mathbb{R}^N : \mu_3'^\top x \leq \mu_3'^\top \mu_1, x \succeq 0\}, \tag{6.30}$$

proving the aforementioned claim. The vectors $\mu_1, \mu_2, \mu_3, \mu_3', \omega_r[k+1]$, together with the convex polytope $\mathcal{P}(\omega_r[k])$, are illustrated in Fig. 6.2c for the case of $N = 2$. Notice from (6.30) that $\mathcal{P}(\omega_r[k])$ may be empty, and that it is empty if and only if $\mu_3'^\top \mu_1 < 0$.

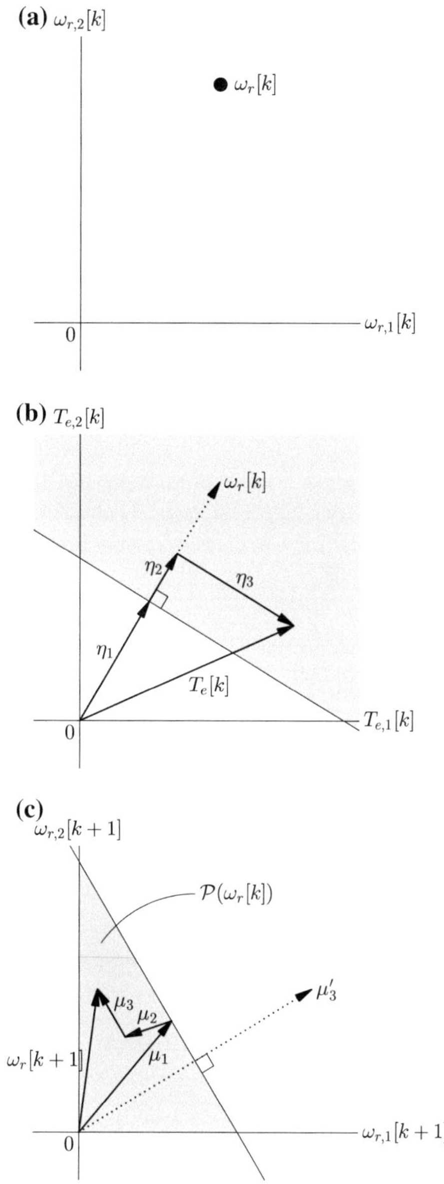

Fig. 6.2 Illustration of the idea behind the numerical method using the case of $N = 2$. $\omega_r[k]$ space (**a**). $T_e[k]$ space (**b**). $\omega_r[k+1]$ space (**c**)

The above analysis is valid for any time $k \in \{0, 1, 2, \ldots\}$ and any state $\omega_r[k] \succeq 0$. Therefore, for any initial state $\omega_r[0] \succeq 0$, upon satisfying (6.25)–(6.27) for $k = 0$, the ensuing state $\omega_r[1]$ must be in $\mathcal{P}(\omega_r[0])$. Similarly, for any ensuing state $\omega_r[1] \in \mathcal{P}(\omega_r[0])$, upon satisfying (6.25)–(6.27) for $k = 1$, the state $\omega_r[2]$ must be in $\mathcal{P}(\omega_r[1])$. It follows that for any initial state $\omega_r[0] \succeq 0$, upon satisfying (6.25)–(6.27) for *both* $k = 0$ and $k = 1$, the state $\omega_r[2]$ must be in $\cup_{\omega_r[1] \in \mathcal{P}(\omega_r[0])} \mathcal{P}(\omega_r[1])$, which can also be written as $\mathcal{P}(\mathcal{P}(\omega_r[0]))$. Continuing in this fashion, we can say that for any time $\ell \in \{1, 2, \ldots\}$ and any initial state $\omega_r[0] \succeq 0$, upon satisfying (6.25)–(6.27) for $k = 0, 1, \ldots, \ell - 1$, the state $\omega_r[\ell]$ must be such that

$$\omega_r[\ell] \in \mathcal{R}(\ell, \omega_r[0]) \triangleq \underbrace{\mathcal{P}\Big(\mathcal{P}\big(\cdots \mathcal{P}}_{\ell \text{ times}}(\omega_r[0]) \cdots\big)\Big), \tag{6.31}$$

where $\mathcal{R}(\ell, \omega_r[0]) \subset \mathbb{R}^N$ defined in (6.31) may be referred to as the *reachable region at time* ℓ *with initial state* $\omega_r[0]$—much like the notion of reachability in control theory. Expression (6.31) also suggests that the sequence of reachable regions $\mathcal{R}(\ell, \omega_r[0])$ for $\ell \in \{1, 2, \ldots\}$ may be computed recursively, i.e.,

$$\mathcal{R}(\ell, \omega_r[0]) = \mathcal{P}\big(\mathcal{R}(\ell - 1, \omega_r[0])\big) \quad \text{with} \quad \mathcal{R}(0, \omega_r[0]) = \{\omega_r[0]\}, \tag{6.32}$$

where each step of the computation uses the fact that $\mathcal{P}(\cdot)$ maps a point (e.g., the one in Fig. 6.2a) to either an empty set or a simple convex polytope (e.g., the one in Fig. 6.2c). These reachable regions can then be used to determine whether a quadruplet $(\omega_r^0, T, \omega_r^T, P_{d,wf})$ is achievable:

$$(\omega_r^0, T, \omega_r^T, P_{d,wf}) \in \mathcal{A} \quad \text{if and only if} \quad \omega_r^T \in \mathcal{R}(T/T_s, \omega_r^0), \tag{6.33}$$

where $T/T_s \in \{1, 2, \ldots\}$. Thus, (6.30), (6.32), and (6.33) together describe a numerical method for characterizing the set $\mathcal{A}$.

6.6 Illustrative Examples

When the wind farm has exactly $N = 2$ wind turbines, the sequence of reachable regions $\mathcal{R}(k, \omega_r[0])$ for $k \in \{1, 2, \ldots\}$ can be graphically visualized on a plane. To illustrate this visualization, consider the two examples portrayed in Fig. 6.3a, b.

Figure 6.3a shows the result of executing the aforementioned numerical method for a wind farm with $N = 2$ wind turbines and a constant desired power output of $\overline{P}_{d,wf} = 0.9$, which is *below* the wind farm capacity of 1. In Fig. 6.3a, subplot (1, 1) shows the initial state $(\omega_{r,1}[0], \omega_{r,2}[0]) = (0.5, 0.5)$ as a black dot. Subplot (1, 2) shows in gray the set of possible $(\omega_{r,1}[1], \omega_{r,2}[1])$ upon satisfying the dynamics (6.25), state constraint (6.26), and grid operator requirement (6.27) for $k = 0$. As defined earlier in (6.31), this gray region is the reachable region at time $k = 1$.

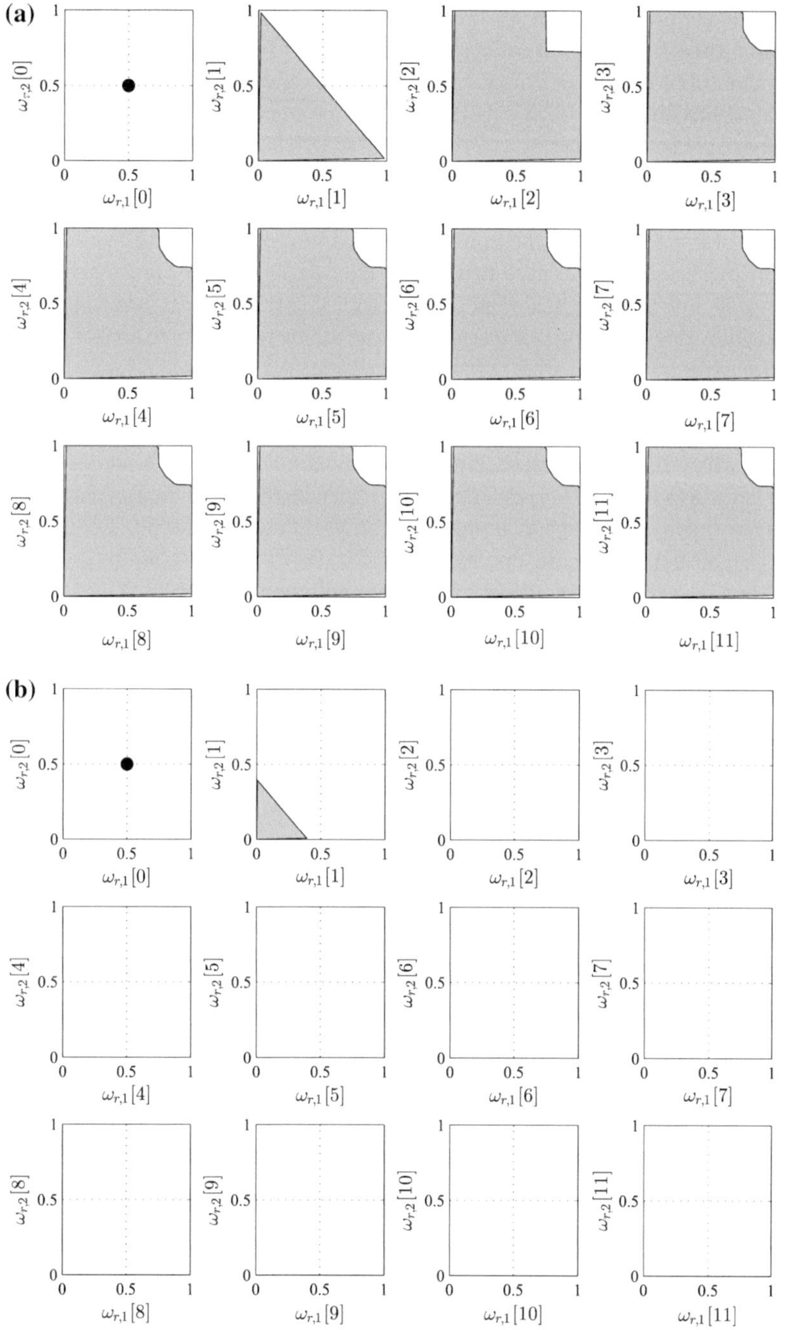

Fig. 6.3 Application of the numerical method to a wind farm with $N = 2$ wind turbines, in which the reachable regions are shaded in *gray*. Desired power output $\overline{P}_{d,wf} = 0.9$ is *below* wind farm capacity of 1 (**a**). Desired power output $\overline{P}_{d,wf} = 1.05$ is *above* wind farm capacity of 1 (**b**)

Similarly, subplot $(1, 3)$ shows the set of possible $(\omega_{r,1}[2], \omega_{r,2}[2])$, or reachable region at time $k = 2$, upon satisfying (6.25)–(6.27) for both $k = 0$ and $k = 1$. In general, a subplot with horizontal and vertical axes labeled $\omega_{r,1}[k]$ and $\omega_{r,2}[k]$ shows the set of possible $(\omega_{r,1}[k], \omega_{r,2}[k])$, or reachable region at time k, upon satisfying (6.25)–(6.27) for time $0, 1, \ldots, k-1$. Observe that as time k elapses, the sequence of reachable regions converge to a steady-state shape and remain nonempty. This suggests that with the initial state $(\omega_{r,1}[0], \omega_{r,2}[0]) = (0.5, 0.5)$, it is possible for the wind farm to produce the constant desired power output of $\overline{P}_{d,wf} = 0.9$ in a sustainable fashion. Also note that points that are not in the gray regions are not achievable. For instance, the point $(\omega_{r,1}[11], \omega_{r,2}[11]) = (0.9, 0.9)$ in subplot $(3, 4)$ is not achievable, meaning that it is not possible for the wind farm to go from the initial state $(\omega_{r,1}[0], \omega_{r,2}[0]) = (0.5, 0.5)$ to the final state $(\omega_{r,1}[11], \omega_{r,2}[11]) = (0.9, 0.9)$ and satisfy (6.25)–(6.27) along the way.

Using the same format as Fig. 6.3a, b shows the result of executing the numerical method when the constant desired power output is $\overline{P}_{d,wf} = 1.05$, which is *above* the wind farm capacity of 1. Observe that in this "unsustainable" scenario, the wind farm is able to satisfy the state constraint (6.26) and grid operator requirement (6.27) for $k = 0$ because of the kinetic energy stored in it. However, its state $(\omega_{r,1}[1], \omega_{r,2}[1])$ must then be in the gray triangle (i.e., convex polytope) in subplot $(1, 2)$, signifying that its two turbines have slowed down. Also observe that the wind farm is *not* able to satisfy (6.26) and (6.27) for both $k = 0$ and $k = 1$ because it has run out of kinetic energy, which can be deduced by noting that the reachable region at time $k = 2$ is empty. Indeed, the same can be said for all $k \geq 2$. Therefore, when the desired power output exceeds the wind farm capacity, the reachable region essentially disappears after $k = 2$ time instants. Note that Fig. 6.3a, b allow us to graphically visualize what can be achieved and what cannot be, when it comes to kinetic energy release.

6.7 Concluding Remarks

In this chapter, we have addressed from a theoretical viewpoint the emerging topic of controlling the kinetic energy stored in wind farms as a means to maintain transient frequency stability. We have introduced the concept of achievability, which indicates whether a given quadruplet of initial state, final time, final state, and desired wind farm power output is achievable by releasing or storing kinetic energy in the wind farm. We have also discussed the practical implications of achievability, presented an analytical method and a numerical method for characterizing it, and illustrated the numerical method via two examples.

We stress that this chapter is not about designing a specific wind farm controller for kinetic energy release, but rather about understanding the fundamental limit of such controllers. We also note that the results of this chapter may be extended in a number of directions as open challenges for future research. For instance, the definition of achievability may be modified to incorporate additional constraints that arise in a real wind farm. As another example, the question of achievability may be framed in a more general context of a wind farm plus a photovoltaic system, or a wind farm plus a battery storage, or other regular combinations. As yet another example, when a given quadruplet happens to be achievable, one may wish to find the "best" control input that optimizes some cost function subject to realizing the quadruplet, in which case optimal control theory can be used to gain insights into the nature of the optimal strategy. Such insights can then be used to design practical, near-optimal wind farm controllers for kinetic energy release.

Chapter 7
Conclusion

In this monograph, we have addressed a number of major issues that arise in the control and operation of grid-connected wind farms. In particular, we have shown how systems and control theory may be applied to develop contemporary solutions to these important issues.

As a summary, in Chap. 2, we have described a mathematical model for variable-speed wind turbines employing DFIGs. Based on this model, we have utilized a number of control techniques—including feedback linearization, model order reduction, uncertainty estimation, and potential function minimization—to design a reconfigurable nonlinear dual-mode controller, which enables two most desirable operating modes of MPT and PR, while addressing other issues such as uncertainties in most of the model parameters. We have also demonstrated the effectiveness of this single-turbine controller through simulation with a realistic wind profile.

In Chap. 3, we have presented a simple approximate model, which tries to mimic the active and reactive power dynamics of generic analytical and empirical WTCS models, along with two parameter identification schemes, which determine the approximate model parameters in both cases. We have also demonstrated through simulation the ability of the approximate model in resembling several different analytical and empirical WTCS models from the literature and from real data. The results suggest that the approximate model is a compelling candidate, based on which one may design and analyze a second-to-minute-timescale supervisory WFC using a variety of control techniques.

In Chap. 4, we have developed a novel supervisory WFC that enables the active power output of a wind farm to accurately and smoothly track a desired reference provided by a power grid operator. In the process of developing this WFC, we have explained the rationale behind its two-loop architecture. We have also shown that the outer feedback loop of the WFC contains a model predictive controller, which uses various forecasts and feedbacks to iteratively compute a set of desired power trajectories, so that the deterministic tracking accuracy of the wind farm power output on a receding horizon is optimized. In contrast, the inner feedback loop contains an adaptive controller, which uses estimated wind speed characteristics to adaptively tune a set of proportional controller gains, so that the stochastic smoothness of the

J.N. Jiang et al., *Control and Operation of Grid-Connected Wind Farms*,
Advances in Industrial Control, DOI 10.1007/978-3-319-39135-9_7

wind farm power output on a shorter timescale is optimized. Lastly, we have carried out a series of simulation studies that demonstrate the effectiveness of the WFC, including its ability to exploit forecast availability, design freedom, and wind speed correlation.

In Chap. 5, we have applied the recently-developed theory of QLC to the design of proportional controller gains on the inner feedback loop of the WFC from Chap. 4. By taking the asymmetric saturation nonlinearities directly into account at the design stage, QLC leads to controller gains that are superior in performance compared to those designed by traditional linear methods which neglect saturation. Indeed, the performance improvements are found to be significant—up to 60 % reduction in the values of the cost function—over a broad range of wind farm operating regimes, including low and high wind speed, medium and high power generation, weak and strong wind correlation, and cheap and expensive control regimes.

In Chap. 6, we have addressed from a theoretical viewpoint the emerging topic of controlling the kinetic energy stored in wind farms as a means to maintain transient frequency stability. We have introduced the concept of achievability, which indicates whether a given quadruplet of initial state, final time, final state, and desired wind farm power output is achievable by releasing or storing kinetic energy in the wind farm. We have also discussed the practical implications of achievability, presented an analytical method and a numerical method for characterizing it, and illustrated the numerical method via two examples. We note that this chapter is not about designing a specific WFC for kinetic energy release, but rather about understanding the fundamental limit of such controllers.

As a closing remark, we believe that control and operation of grid-connected wind farms will remain a vibrant area of research for years to come, and we hope that the ideas contained in this monograph will make a difference in the area.

References

1. Interim report—system disturbance on 4 November 2006. Executive Summary, Union for the Co-Ordination of Transmission of Electricity, Brussels, Belgium (2006)
2. 20% wind energy by 2030—increasing wind energy's contribution to U.S. electricity supply. Executive Summary, U.S. Department of Energy, Washington, DC (2008)
3. Standard models for variable generation. Special Report, North American Electric Reliability Corporation, Princeton, NJ (2010)
4. NREL leads wind farm modeling research. Continuum Magazine, Issue 3, National Renewable Energy Laboratory, Golden, CO (2012)
5. The NREL spectrum of clean energy innovation. Continuum Magazine, Issue 3, National Renewable Energy Laboratory, Golden, CO (2012)
6. J. Aho, A. Buckspan, J. Laks, P. Fleming, Y. Jeong, F. Dunne, M. Churchfield, L. Pao, K. Johnson, A tutorial of wind turbine control for supporting grid frequency through active power control, in *Proceedings of the American Control Conference*, Montreal, Canada (2012), pp. 3120–3131
7. M. Akbari, S.M. Madani, Analytical evaluation of control strategies for participation of doubly fed induction generator-based wind farms in power system short-term frequency regulation. IET Renew. Power Gener. **8**(3), 324–333 (2014)
8. M. Asmine, J. Brochu, J. Fortmann, R. Gagnon, Y. Kazachkov, C.-E. Langlois, C. Larose, E. Muljadi, J. MacDowell, P. Pourbeik, S.A. Seman, K. Wiens, Model validation for wind turbine generator models. IEEE Trans. Power Syst. **26**(3), 1769–1782 (2011)
9. K.J. Astrom, B. Wittenmark, *Adaptive Control*, 2nd edn. (Prentice Hall, Upper Saddle River, 1994)
10. A.B.T. Attya, T. Hartkopf, Control and quantification of kinetic energy released by wind farms during power system frequency drops. IET Renew. Power Gener. **7**(3), 210–224 (2013)
11. A.B.T. Attya, T. Hartkopf, Wind turbine contribution in frequency drop mitigation - modified operation and estimating released supportive energy. IET Gener. Transm. Distrib. **8**(5), 862–872 (2014)
12. B. Beltran, T. Ahmed-Ali, M.E.H. Benbouzid, Sliding mode power control of variable-speed wind energy conversion systems. IEEE Trans. Energy Convers. **23**(2), 551–558 (2008)
13. D.P. Bertsekas, *Nonlinear Programming*, 2nd edn. (Athena Scientific, Belmont, 1999)
14. F.D. Bianchi, H. De Battista, R.J. Mantz, *Wind Turbine Control Systems: Principles, Modelling and Gain Scheduling Design* (Springer, London, 2007)
15. B.K. Bose, *Modern Power Electronics and AC Drives* (Prentice Hall, Upper Saddle River, 2002)
16. S. Boyd, L. Vandenberghe, *Convex Optimization* (Cambridge University Press, New York, 2004)

J.N. Jiang et al., *Control and Operation of Grid-Connected Wind Farms*,
Advances in Industrial Control, DOI 10.1007/978-3-319-39135-9

17. W.L. Brogan, *Modern Control Theory*, 3rd edn. (Prentice Hall, Englewood Cliffs, 1991)
18. B.G. Brown, R.W. Katz, A.H. Murphy, Time series models to simulate and forecast wind speed and wind power. J. Clim. Appl. Meteorol. **23**(8), 1184–1195 (1984)
19. A. Buckspan, J. Aho, P. Fleming, Y. Jeong, L. Pao, Combining droop curve concepts with control systems for wind turbine active power control, in *Proceedings of the IEEE Symposium on Power Electronics and Machines in Wind Applications*, Denver, CO (2012), pp. 1–8
20. A. Buckspan, L. Pao, J. Aho, P. Fleming, Stability analysis of a wind turbine active power control system, in *Proceedings of the American Control Conference*, Washington, DC (2013), pp. 1418–1423
21. E.F. Camacho, C. Bordons, *Model Predictive Control*, 2nd edn. (Springer, London, 2004)
22. H.R. Chamorro, M. Ghandhari, R. Eriksson, Wind power impact on power system frequency response, in *Proceedings of the North American Power Symposium*, Manhattan, KS (2013), pp. 1–6
23. L.-R. Chang-Chien, W.-T. Lin, Y.-C. Yin, Enhancing frequency response control by DFIGs in the high wind penetrated power systems. IEEE Trans. Power Syst. **26**(2), 710–718 (2011)
24. L.-R. Chang-Chien, C.-C. Sun, Y.-J. Yeh, Modeling of wind farm participation in AGC. IEEE Trans. Power Syst. **29**(3), 1204–1211 (2014)
25. L.-R. Chang-Chien, Y.-C. Yin, Strategies for operating wind power in a similar manner of conventional power plant. IEEE Trans. Energy Convers. **24**(4), 926–934 (2009)
26. N.R. Chaudhuri, B. Chaudhuri, Considerations toward coordinated control of DFIG-based wind farms. IEEE Trans. Power Deliv. **28**(3), 1263–1270 (2013)
27. R.B. Chedid, S.H. Karaki, C. El-Chamali, Adaptive fuzzy control for wind-diesel weak power systems. IEEE Trans. Energy Convers. **15**(1), 71–78 (2000)
28. C.-T. Chen, *Linear System Theory and Design*, 3rd edn. (Oxford University Press, New York, 1999)
29. S. Ching, Y. Eun, C. Gokcek, P.T. Kabamba, S.M. Meerkov, *Quasilinear Control* (Cambridge University Press, New York, 2011)
30. K. Clark, N.W. Miller, J.J. Sanchez-Gasca, *Modeling of GE Wind Turbine-Generators for Grid Studies. Version 4.2* (General Electric International, Inc., Schenectady, 2008)
31. J. Creaby, Y. Li, J.E. Seem, Maximizing wind turbine energy capture using multivariable extremum seeking control. Wind Eng. **33**(4), 361–388 (2009)
32. R.G. de Almeida, E.D. Castronuovo, J.A. Pecas Lopes, Optimum generation control in wind parks when carrying out system operator requests. IEEE Trans. Power Syst. **21**(2), 718–725 (2006)
33. R.G. de Almeida, J.A. Pecas Lopes, Participation of doubly fed induction wind generators in system frequency regulation. IEEE Trans. Power Syst. **22**(3), 944–950 (2007)
34. G. Delille, B. François, G. Malarange, Dynamic frequency control support by energy storage to reduce the impact of wind and solar generation on isolated power system's inertia. IEEE Trans. Sustain. Energy **3**(4), 931–939 (2012)
35. A.S. Dobakhshari, M. Fotuhi-Firuzabad, A reliability model of large wind farms for power system adequacy studies. IEEE Trans. Energy Convers. **24**(3), 792–801 (2009)
36. J. Ekanayake, N. Jenkins, Comparison of the response of doubly fed and fixed-speed induction generator wind turbines to changes in network frequency. IEEE Trans. Energy Convers. **19**(4), 800–802 (2004)
37. E. Ela, V. Gevorgian, P. Fleming, Y.C. Zhang, M. Singh, E. Muljadi, A. Scholbrook, J. Aho, A. Buckspan, L. Pao, V. Singhvi, A. Tuohy, P. Pourbeik, D. Brooks, N. Bhatt, Active power controls from wind power: bridging the gaps. Technical Report, National Renewable Energy Laboratory, Golden, CO (2014)
38. R. Fadaeinedjad, M. Moallem, G. Moschopoulos, Simulation of a wind turbine with doubly-fed induction generator by FAST and Simulink. IEEE Trans. Energy Convers. **23**(2), 690–700 (2008)
39. A.E. Feijoo, J. Cidras, Modeling of wind farms in the load flow analysis. IEEE Trans. Power Syst. **15**(1), 110–115 (2000)

40. L.M. Fernandez, C.A. Garcia, F. Jurado, Comparative study on the performance of control systems for doubly fed induction generator (DFIG) wind turbines operating with power regulation. Energy **33**(9), 1438–1452 (2008)
41. S. Frandsen, R. Barthelmie, S. Pryor, O. Rathmann, S. Larsen, J. Hojstrup, M. Thogersen, Analytical modelling of wind speed deficit in large offshore wind farms. Wind Energy **9**(1–2), 39–53 (2006)
42. V. Galdi, A. Piccolo, P. Siano, Designing an adaptive fuzzy controller for maximum wind energy extraction. IEEE Trans. Energy Convers. **23**(2), 559–569 (2008)
43. H. Geng, G. Yang, Robust pitch controller for output power levelling of variable-speed variable-pitch wind turbine generator systems. IET Renew. Power Gener. **3**(2), 168–179 (2009)
44. E.M. Gertz, S.J. Wright, Object-oriented software for quadratic programming. ACM Trans. Math. Softw. **29**(1), 58–81 (2003)
45. G. Giebel, R. Brownsword, G. Kariniotakis, M. Denhard, C. Draxl, The state-of-the-art in short-term prediction of wind power: a literature overview, 2nd edn. Annual Report, Riso National Laboratory, Roskilde, Denmark (2011)
46. F. Gonzalez-Longatt, E. Chikuni, W. Stemmet, K. Folly, Effects of the synthetic inertia from wind power on the total system inertia after a frequency disturbance, in *Proceedings of the IEEE Power Engineering Society Conference and Exposition in Africa*, Johannesburg, South Africa (2012), pp. 1–7
47. I.A. Gowaid, A. El-Zawawi, M. El-Gammal, Improved inertia and frequency support from grid-connected DFIG wind farms, in *Proceedings of the IEEE Power Systems Conference and Exposition*, Phoenix, AZ (2011), pp. 1–9
48. Y. Guo, S.H. Hosseini, J.N. Jiang, C.Y. Tang, R.G. Ramakumar, Voltage/pitch control for maximisation and regulation of active/reactive powers in wind turbines with uncertainties. IET Renew. Power Gener. **6**(2), 99–109 (2012)
49. A.D. Hansen, P. Sorensen, F. Iov, F. Blaabjerg, Control of variable speed wind turbines with doubly-fed induction generators. Wind Eng. **28**(4), 411–434 (2004)
50. A.D. Hansen, P. Sorensen, F. Iov, F. Blaabjerg, Centralised power control of wind farm with doubly fed induction generators. Renew. Energy **31**(7), 935–951 (2006)
51. S. Heier, *Grid Integration of Wind Energy Conversion Systems* (Wiley, New York, 1998)
52. B. Hopfensperger, D.J. Atkinson, R.A. Lakin, Stator-flux-oriented control of a doubly-fed induction machine with and without position encoder. IEE Proc. Electr. Power Appl. **147**(4), 241–250 (2000)
53. http://www.ampl.com
54. http://www.cvxopt.org
55. http://www.ict-aeolus.eu
56. http://www.QuasilinearControl.com
57. E. Iyasere, M. Salah, D. Dawson, J. Wagner, Nonlinear robust control to maximize energy capture in a variable speed wind turbine, in *Proceedings of the American Control Conference*, Seattle, WA (2008), pp. 1824–1829
58. K.E. Johnson, L.J. Fingersh, M.J. Balas, L.Y. Pao, Methods for increasing region 2 power capture on a variable-speed wind turbine. J. Solar Energy Eng. **126**(4), 1092–1100 (2004)
59. K.E. Johnson, L.Y. Pao, M.J. Balas, L.J. Fingersh, Control of variable-speed wind turbines: standard and adaptive techniques for maximizing energy capture. IEEE Control Syst. Mag. **26**(3), 70–81 (2006)
60. B.R. Karthikeya, R.J. Schütt, Overview of wind park control strategies. IEEE Trans. Sustain. Energy **5**(2), 416–422 (2014)
61. P.-K. Keung, P. Li, H. Banakar, B.T. Ooi, Kinetic energy of wind-turbine generators for system frequency support. IEEE Trans. Power Syst. **24**(1), 279–287 (2009)
62. H.K. Khalil, *Nonlinear Systems*, 3rd edn. (Prentice Hall, Upper Saddle River, 2002)
63. T. Knudsen, T. Bak, M. Soltani, Distributed control of large-scale offshore wind farms, in *Proceedings of the European Wind Energy Conference and Exhibition*, Marseille, France (2009

64. T. Knudsen, T. Bak, M. Soltani, Prediction models for wind speed at turbine locations in a wind farm. Wind Energy **14**(7), 877–894 (2011)
65. H.-S. Ko, G.-G. Yoon, W.-P. Hong, Active use of DFIG-based variable-speed wind-turbine for voltage regulation at a remote location. IEEE Trans. Power Syst. **22**(4), 1916–1925 (2007)
66. P.C. Krause, O. Wasynczuk, S.D. Sudhoff, *Analysis of Electric Machinery and Drive Systems*, 2nd edn. (Wiley-IEEE Press, Piscataway, 2002)
67. H. Kwakernaak, R. Sivan, *Linear Optimal Control Systems* (Wiley-InterScience, New York, 1972)
68. Y. Lei, A. Mullane, G. Lightbody, R. Yacamini, Modeling of the wind turbine with a doubly fed induction generator for grid integration studies. IEEE Trans. Energy Convers. **21**(1), 257–264 (2006)
69. H. Li, K.L. Shi, P.G. McLaren, Neural-network-based sensorless maximum wind energy capture with compensated power coefficient. IEEE Trans. Ind. Appl. **41**(6), 1548–1556 (2005)
70. P. Li, P.-K. Keung, B.-T. Ooi, Development and simulation of dynamic control strategies for wind farms. IET Renew. Power Gener. **3**(2), 180–189 (2009)
71. L. Ljung, *System Identification: Theory for the User*, 2nd edn. (Prentice Hall, Upper Saddle River, 1999)
72. Z. Lubosny, J.W. Bialek, Supervisory control of a wind farm. IEEE Trans. Power Syst. **22**(3), 985–994 (2007)
73. H.T. Ma, B.H. Chowdhury, Working towards frequency regulation with wind plants: combined control approaches. IET Renew. Power Gener. **4**(4), 308–316 (2010)
74. J.R. Marden, S.D. Ruben, L. Pao, Surveying game theoretic approaches for wind farm optimization, in *Proceedings of the AIAA Aerospace Sciences Meeting*, Nashville, TN (2012), pp. 1–10
75. B. Marinescu, A robust coordinated control of the doubly-fed induction machine for wind turbines: a state-space based approach, in *Proceedings of the American Control Conference*, Boston, MA (2004), pp. 174–179
76. N.W. Miller, J.J. Sanchez-Gasca, W.W. Price, R.W. Delmerico, Dynamic modeling of GE 1.5 and 3.6 MW wind turbine-generators for stability simulations, in *Proceedings of the IEEE Power and Energy Society General Meeting*, Toronto, Canada (2003), pp. 1977–1983
77. A. Monroy, L. Alvarez-Icaza, G. Espinosa-Perez, Passivity-based control for variable speed constant frequency operation of a DFIG wind turbine. Int. J. Control **81**(9), 1399–1407 (2008)
78. C. Monteiro, R. Bessa, V. Miranda, A. Botterud, J. Wang, G. Conzelmann, Wind power forecasting: state-of-the-art 2009. Technical Report, Argonne National Laboratory, Argonne, IL (2009)
79. J. Morren, S.W.H. de Haan, W.L. Kling, J.A. Ferreira, Wind turbines emulating inertia and supporting primary frequency control. IEEE Trans. Power Syst. **21**(1), 433–434 (2006)
80. J. Morren, J. Pierik, S.W.H. de Haan, Inertial response of variable speed wind turbines. Electric Power Syst. Res. **76**(11), 980–987 (2006)
81. C.F. Moyano, J.A. Pecas Lopes, An optimization approach for wind turbine commitment and dispatch in a wind park. Electric Power Syst. Res. **79**(1), 71–79 (2009)
82. E. Muljadi, C.P. Butterfield, Pitch-controlled variable-speed wind turbine generation. IEEE Trans. Ind. Appl. **37**(1), 240–246 (2001)
83. E. Muljadi, M. Singh, V. Gevorgian, Fixed-speed and variable-slip wind turbines providing spinning reserves to the grid, in *Proceedings of the IEEE Power and Energy Society General Meeting*, Vancouver, Canada (2013), pp. 1–5
84. K. Narendra, P. Gallman, An iterative method for the identification of nonlinear systems using a Hammerstein model. IEEE Trans. Autom. Control **11**(3), 546–550 (1966)
85. H.R. Ossareh, Quasilinear control theory for systems with asymmetric actuators and sensors. Ph.D. Dissertation, University of Michigan, Ann Arbor, MI (2013)
86. L.Y. Pao, K.E. Johnson, Control of wind turbines. IEEE Control Syst. Mag. **31**(2), 44–62 (2011)
87. R. Pena, J.C. Clare, G.M. Asher, Doubly fed induction generator using back-to-back PWM converters and its application to variable-speed wind-energy generation. IEE Proc. Electr. Power Appl. **143**(3), 231–241 (1996)

88. S. Peresada, A. Tilli, A. Tonielli, Power control of a doubly fed induction machine via output feedback. Control Eng. Pract. **12**(1), 41–57 (2004)
89. W. Qiao, W. Zhou, J.M. Aller, R.G. Harley, Wind speed estimation based sensorless output maximization control for a wind turbine driving a DFIG. IEEE Trans. Power Electron. **23**(3), 1156–1169 (2008)
90. B.G. Rawn, M. Gibescu, W.L. Kling, A static analysis method to determine the availability of kinetic energy from wind turbines, in *Proceedings of the IEEE Power and Energy Society General Meeting*, Minneapolis, MN (2010), pp. 1–8
91. E. Rezaei, A. Tabesh, M. Ebrahimi, Dynamic model and control of DFIG wind energy systems based on power transfer matrix. IEEE Trans. Power Deliv. **27**(3), 1485–1493 (2012)
92. J.B. Roberts, P.D. Spanos, *Random Vibration and Statistical Linearization* (Dover Publications, Mineola, 2003)
93. J.L. Rodriguez-Amenedo, S. Arnalte, J.C. Burgos, Automatic generation control of a wind farm with variable speed wind turbines. IEEE Trans. Energy Convers. **17**(2), 279–284 (2002)
94. J.A. Rossiter, *Model-Based Predictive Control: A Practical Approach* (CRC Press, Boca Raton, 2003)
95. T. Senjyu, R. Sakamoto, N. Urasaki, T. Funabashi, H. Fujita, H. Sekine, Output power leveling of wind turbine generator for all operating regions by pitch angle control. IEEE Trans. Energy Convers. **21**(2), 467–475 (2006)
96. T. Senjyu, R. Sakamoto, N. Urasaki, T. Funabashi, H. Sekine, Output power leveling of wind farm using pitch angle control with fuzzy neural network, in *Proceedings of the IEEE Power and Energy Society General Meeting*, Montreal, Canada (2006), pp. 1–8
97. S. Sharma, S.-H. Huang, N.D.R. Sarma, System inertial frequency response estimation and impact of renewable resources in ERCOT interconnection, in *Proceedings of the IEEE Power and Energy Society General Meeting*, San Diego, CA (2011), pp. 1–6
98. J. Shu, B.H. Zhang, Z.Q. Bo, A wind farm coordinated controller for power optimization, in *Proceedings of the IEEE Power and Energy Society General Meeting*, San Diego, CA (2011), pp. 1–8
99. M. Singh, S. Santoso, Dynamic models for wind turbines and wind power plants. Subcontract Report, National Renewable Energy Laboratory, Golden, CO (2011)
100. M. Soleimanzadeh, R. Wisniewski, Wind speed dynamical model in a wind farm, in *Proceedings of the IEEE International Conference on Control and Automation*, Xiamen, China (2010), pp. 2246–2250
101. M. Soleimanzadeh, R. Wisniewski, Controller design for a wind farm, considering both power and load aspects. Mechatronics **21**(4), 720–727 (2011)
102. M. Soltani, T. Knudsen, T. Bak, Modeling and simulation of offshore wind farms for farm level control, in *Proceedings of the European Offshore Wind Conference and Exhibition*, Stockholm, Sweden (2009)
103. V. Spudic, M. Jelavic, M. Baotic, Wind turbine power references in coordinated control of wind farms. Autom.-J. Control Meas. Electron. Comput. Commun. **52**(2), 82–94 (2011)
104. V. Spudic, M. Jelavic, M. Baotic, N. Peric, Hierarchical wind farm control for power/load optimization, in *Proceedings of the TORQUE 2010: The Science of Making Torque from Wind*, Crete, Greece (2010), pp. 681–692
105. K. Stol, M.J. Balas, Full-state feedback control of a variable-speed wind turbine: a comparison of periodic and constant gains. J. Solar Energy Eng. **123**(4), 319–326 (2001)
106. C.Y. Tang, Y. Guo, J.N. Jiang, Nonlinear dual-mode control of variable-speed wind turbines with doubly fed induction generators. IEEE Trans. Control Syst. Technol. **19**(4), 744–756 (2011)
107. G. Tapia, A. Tapia, J.X. Ostolaza, Proportional-integral regulator-based approach to wind farm reactive power management for secondary voltage control. IEEE Trans. Energy Convers. **22**(2), 488–498 (2007)
108. G.C. Tarnowski, R. Reginatto, Adding active power regulation to wind farms with variable speed induction generators, in *Proceedings of the IEEE Power and Energy Society General Meeting*, Tampa, FL (2007), pp. 1–8

109. M. Tsili, S. Papathanassiou, A review of grid code technical requirements for wind farms. IET Renew. Power Gener. **3**(3), 308–332 (2009)
110. K.V. Vidyanandan, N. Senroy, Primary frequency regulation by deloaded wind turbines using variable droop. IEEE Trans. Power Syst. **28**(2), 837–846 (2013)
111. E. Vittal, A. Keane, Identification of critical wind farm locations for improved stability and system planning. IEEE Trans. Power Syst. **28**(3), 2950–2958 (2013)
112. A.D. Wright, M.J. Balas, Design of state-space-based control algorithms for wind turbine speed regulation. J. Solar Energy Eng. **125**(4), 386–395 (2003)
113. F. Wu, X. Zhang, P. Ju, M.J.H. Sterling, Decentralized nonlinear control of wind turbine with doubly fed induction generator. IEEE Trans. Power Syst. **23**(2), 613–621 (2008)
114. D. Zhi, L. Xu, Direct power control of DFIG with constant switching frequency and improved transient performance. IEEE Trans. Energy Convers. **22**(1), 110–118 (2007)

Index

J.N. Jiang et al., *Control and Operation of Grid-Connected Wind Farms*,
Advances in Industrial Control, DOI 10.1007/978-3-319-39135-9

MIX
Papier aus verantwortungsvollen Quellen
Paper from responsible sources
FSC® C105338

If you have any concerns about our products,
you can contact us on
ProductSafety@springernature.com

In case Publisher is established outside the EU,
the EU authorized representative is:
Springer Nature Customer Service Center GmbH
Europaplatz 3, 69115 Heidelberg, Germany

Printed by Libri Plureos GmbH
in Hamburg, Germany